Know This

Also by John Brockman

As Author
By the Late John Brockman
37
Afterwords
The Third Culture: Beyond the Scientific Revolution
Digerati

As Editor
About Bateson
Speculations
Doing Science
Ways of Knowing
Creativity
The Greatest Innovations of the Past 2,000 Years
The Next Fifty Years
The New Humanists
Curious Minds
What We Believe but Cannot Prove
My Einstein
Intelligent Thought
What Is Your Dangerous Idea?
What Are You Optimistic About?
What Have You Changed Your Mind About?
This Will Change Everything
Is the Internet Changing the Way You Think?
Culture
The Mind
This Will Make You Smarter
This Explains Everything
Thinking
What Should We Be Worried About?
The Universe
This Idea Must Die
What to Think About Machines That Think
Life

As Coeditor
How Things Are (with Katinka Matson)

Know This

Today's Most Interesting
and Important Scientific
Ideas, Discoveries, and
Developments

EDITED BY JOHN BROCKMAN

HARPER PERENNIAL

NEW YORK • LONDON • TORONTO • SYDNEY • NEW DELHI • AUCKLAND

HarperCollins books may be purchased for educational, business, or sales promotional use. For information, please email the Special Markets Department at SPsales@harpercollins.com.

FIRST EDITION

Library of Congress Cataloging-in-Publication Data has been applied for.

ISBN 978-0-06-256206-7 (pbk.)

17 18 19 20 21   LSC   10 9 8 7 6 5 4 3 2

For Juliet and Miles

CONTENTS

ACKNOWLEDGMENTS

My thanks to Peter Hubbard of HarperCollins and my agent Max Brockman, for their continued encouragement. A special thanks, once again, to Sara Lippincott for her thoughtful attention to the manuscript.

PREFACE

The *Edge* Question

> Scientific topics receiving prominent play in newspapers and
> magazines over the past several years include molecular biology,
> artificial intelligence, artificial life, chaos theory, massive paral-
> lelism, neural nets, the inflationary universe, fractals, complex
> adaptive systems, superstrings, biodiversity, nanotechnology,
> the human genome, expert systems, punctuated equilibrium,
> cellular automata, fuzzy logic, space biospheres, the Gaia hy-
> pothesis, virtual reality, cyberspace, and teraflop machines. . . .
> Unlike previous intellectual pursuits, the achievements of the
> third culture are not the marginal disputes of a quarrelsome
> mandarin class: they will affect the lives of everybody on the
> planet.

You might think that the above list of topics is a preamble for the
Edge Question of 2016, but you would be wrong. It was a central
point in my essay, "The Third Culture," published twenty-five
years ago in the *Los Angeles Times,* September 19, 1991. The essay,
a manifesto, was a collaborative effort, with input from Stephen
Jay Gould, Murray Gell-Mann, Richard Dawkins, Daniel C.
Dennett, Jared Diamond, Stuart Kauffman, and Nicholas Hum-
phrey among other distinguished scientists and thinkers. It pro-
claimed that "the third culture consists of those scientists and

other thinkers in the empirical world who, through their work and expository writing, are taking the place of the traditional intellectual in rendering visible the deeper meanings of our lives, redefining who and what we are," and it continued:

> What traditionally has been called "science" has today become "public culture." Stewart Brand writes that "Science is the only news. When you scan through a newspaper or magazine, all the human interest stuff is the same old he-said-she-said, the politics and economics the same sorry cyclic dramas, . . . even the technology is predictable if you know the science. Human nature doesn't change much; science does, and the change accrues, altering the world irreversibly." We now live in a world in which the rate of change is the biggest change.

Science has thus become a big story, if not *the* big story: News that will stay news.

This is evident by the continued relevance today of the scientific topics in the 1991 essay, all of which were in play before the Web, social media, mobile communications, deep learning, Big Data. Time for an update. . . .

WHAT DO YOU CONSIDER THE MOST INTERESTING RECENT [SCIENTIFIC] NEWS? WHAT MAKES IT IMPORTANT?

The online response this year is just shy of 200 contributions: Here is the news, sifted by those who often make it.

JOHN BROCKMAN
Publisher, *Edge*

Know This

HUMAN PROGRESS QUANTIFIED

STEVEN PINKER

Johnstone Family Professor, Department of Psychology; Harvard University; author, *The Sense of Style*

Human intuition is a notoriously poor guide to reality. A half-century of psychological research has shown that when people try to assess risks or predict the future, their heads are turned by stereotypes, memorable events, vivid scenarios, and moralistic narratives.

Fortunately, as the bugs in human cognition have become common knowledge, the workaround—objective data—has become more prevalent, and in many spheres of life observers are replacing gut feelings with quantitative analysis. Sports have been revolutionized by *Moneyball*, policy by *Nudge*, punditry by *538.com*, forecasting by tournaments and prediction markets, philanthropy by effective altruism, the healing arts by evidence-based medicine.

This is interesting news, and it's scientific news because the diagnosis comes from cognitive science and the cure from data science. But the most interesting news is that the quantification of life has been extended to the biggest question of all: Have we made progress? Have the collective strivings of the human race against entropy and the nastier edges of evolution succeeded in improving the human condition?

Enlightenment thinkers thought this was possible, of course, and in Victorian times progress became a major theme of Anglo-American thought. But since then, Romantic and

1

counter–Enlightenment pessimism have taken over large swaths of intellectual life, stoked by historical disasters such as the World Wars and by post-1960s concerns with anthropogenic problems such as pollution and inequality. Today it's common to read about "faith" in progress (often a "naïve" faith), which is set against a nostalgia for a better past, an assessment of present decline, and a dread of the dystopia to come.

But the cognitive and data revolutions warn us not to base our assessment of *anything* on subjective impressions or cherry-picked incidents. As long as bad things haven't vanished altogether, there will always be enough to fill the news, and people will intuit that the world is falling apart. The only way to circumvent this illusion is to plot the incidence of good and bad things over time. Most people agree that life is better than death, health better than disease, prosperity better than poverty, knowledge better than ignorance, peace better than war, safety better than violence, freedom better than coercion. That gives us a set of yardsticks by which we can measure whether progress has actually occurred.

The interesting news is that the answer is mostly yes. I had the first inkling of this answer when quantitative historians and political scientists responded to my answer to the 2007 *Edge* question ("What Are You Optimistic About?") with data sets showing that the rate of homicides and war deaths had plummeted over time. Since then, I have learned that progress has been tracked by the other yardsticks. Economic historians and development scholars (including Gregory Clark, Angus Deaton, Charles Kenny, and Steven Radelet) have plotted the growth of prosperity in their data-rich books, and the case has been made even more vividly in Web sites with innovative graphics, such as

Hans Rosling's *Gapminder,* Max Roser's *Our World in Data,* and Marian Tupy's *HumanProgress.*

Among the other upward swoops are these. People are living longer and healthier lives, not just in the developed world but globally. A dozen infectious and parasitic diseases are extinct or moribund. Vastly more children are going to school and learning to read. Extreme poverty has fallen worldwide from 85 to 10 percent. Despite local setbacks, the world is more democratic than ever. Women are better educated, marrying later, earning more, and in more positions of power and influence. Racial prejudice and hate crimes have decreased since data were first recorded. The world is even getting smarter: In every country, IQ has been increasing by three points a decade.

Of course, quantified progress consists of a set of empirical findings; it is not a sign of some mystical ascent or utopian trajectory or divine grace. And so we should expect to find some spheres of life that have remained the same, gotten worse, or are altogether unquantifiable (such as the endless number of apocalypses that may be conjured in the imagination). Greenhouse gases accumulate, fresh water diminishes, species go extinct, nuclear arsenals remain.

Yet even here, quantification can change our understanding. "Eco-modernists" such as Stewart Brand, Jesse Ausubel, and Ruth DeFries have shown that many indicators of environmental health have improved over the last half-century, and that there are long-term historical processes—such as the de-carbonization of energy, the dematerialization of consumption, and the minimization of farmland—that can be further encouraged. Tabulators of nuclear weapons have pointed out that no such weapon has been used since Nagasaki, testing has fallen effectively to zero, proliferation has expanded the club only to nine countries (rather than thirty or more, as was predicted in the 1960s), seventeen countries have given up their programs, and the number of weapons (and hence the number of opportunities for thefts and accidents and the number of obstacles to the eventual goal of zero) has been reduced by five-sixths.

What makes all this important? Foremost, quantified progress is a feedback signal for adjusting what we've been doing. The gifts of progress we have enjoyed are the result of institutions and norms that have become entrenched in the last two centuries: reason, science, technology, education, expertise, democracy, regulated markets, and a moral commitment to human rights and human flourishing. As counter–Enlightenment critics have long pointed out, there's no guarantee that these developments would make us better off. Yet now we know that in fact they *have* left us better off. This means that for all the ways in which the world today falls short of utopia, the norms and institutions of modernity have put us on a good track. We should work on improving them, rather than burning them down in the conviction that nothing could be worse than our current decadence and in the vague hope that something better might rise from their ashes.

Also, quantified human progress emboldens us to seek more of it. A common belief among activists is that any optimistic datum must be suppressed lest it lull people into complacency, and instead one must keep up the heat by wailing about ongoing crises and scolding people for being insufficiently terrified. Unfortunately, this can lead to a complementary danger: fatalism. After being told that the poor might always be with us, that the gods will punish our hubris, that nature will avenge our despoliation, and that the clock is inexorably ticking down to a midnight of nuclear holocaust and climatic catastrophe, it's natural to conclude that resistance is futile and we should party while we can. The empowering feature of a graph is that it invites you to identify the forces pushing a curve up or down, and then to apply them to push it farther in the same direction.

DOING MORE WITH LESS

Theoretical physicist; author, *Dreams of Earth and Sky*

One of the scientific heroes of our time is Pieter van Dokkum, professor in the Yale Astronomy Department and author of a recent book, *Dragonflies*. The book is about insects, illustrated with marvelous photographs of dragonflies taken by van Dokkum in their natural habitats. As an astronomer, he works with another kind of dragonfly. The Dragonfly Telephoto Array consists of ten 16-inch refractor telescopes arranged like the compound eye of a dragonfly. The refracting lenses are coated with optical surface layers designed to give them superb sensitivity to faint extended objects in the sky. For faint extended objects, the Dragonfly array is about 10 times more sensitive than the best large telescopes. The Dragonfly is also about 1,000 times cheaper. The ten refractors cost together about $100,000, compared to $100 million for a big telescope.

Dragonfly recently discovered forty-seven "ultra-diffuse" Milky-Way-sized galaxies in the Coma cluster, more than were expected from computer models of galactic evolution. Each galaxy is embedded in a halo of dark matter whose mass can be determined from the observed velocities of the visible stars. The galaxies have about 100 times more dark mass than visible mass, compared with the ratio of 10 to 1 between dark and visible mass in our own galaxy. The Dragonfly observations reveal a universe with an intense fine-structure of dark-matter clumps, much clumpier than the standard theory of Big Bang cosmology had predicted.

So it happens that a cheap small telescope can make a big new discovery about the structure of the universe. If the cost-effectiveness of a telescope is measured by the ratio of scientific output to financial input, Dragonfly wins by a large factor. This story has a moral. The moral is not that we should put all our money into small telescopes. We still need big telescopes and big organizations to do world-class astronomy. The moral is that a modest fraction of the astronomy budget, perhaps as much as a third, should be reserved for small and cheap projects. From time to time a winner like Dragonfly will emerge.

THE "SPECIALNESS" OF HUMANITY

KURT GRAY

Assistant professor of psychology, University of North Carolina,
Chapel Hill

> *"Then the Lord God formed the man of dust from the ground
> and breathed into his nostrils the breath of life, and the man
> became a living creature."*—Genesis 2:7

Humans have always been convinced of our own specialness,
certain that we sit at the center of the universe. Not long ago,
we thought ourselves to be God's favorite creation, placed on
a newly created Earth, which was orbited by all other celestial
bodies. We believed that humans were fundamentally different
from other animals and possessed intelligence that could never
be duplicated. Those ideas made us feel comfortable and safe and
so were easy to believe. But they were wrong.

Copernicus and Galileo revealed that the Sun, not the Earth,
lay at the center of the solar system. Charles Lyell revealed that
the Earth was much older than previously thought. Darwin revealed that humans were not fundamentally different from other
animals. Each of these scientific discoveries challenged our presumed specialness. Of course, even if people were just apes with
large frontal cortices, at least we could claim that humans are
part of a very special club—that of living creatures. We marvel
at the beauty of life, the diversity of plants, animals, insects, and
bacteria. Unfortunately, one recent theory undermines the specialness of all life.

The MIT physicist Jeremy England has suggested that life is merely an inevitable consequence of thermodynamics. He argues that living systems are the best way of dissipating energy from an external source: Bacteria, beetles, and humans are the most efficient way to use up sunlight. According to England, the process of entropy means that molecules that sit long enough under a heat lamp will eventually structure themselves to metabolize, move, and self-replicate—i.e., become alive. Granted, this process might take billions of years, but in this view living creatures are little different from other physical structures that move and replicate with the addition of energy, such as vortices in flowing water (driven by gravity) and sand dunes in the desert (driven by wind). England's theory not only blurs the line between the living and the nonliving but also further undermines the specialness of humanity. It suggests that what humans are especially good at is nothing more than using up energy (something we seem to do with great gusto)—a kind of specialness that hardly lifts our hearts.

J. M. BERGOGLIO'S 2015 REVIEW OF GLOBAL ECOLOGY

STUART PIMM

Doris Duke Professor of Conservation Ecology, Duke University; author, *The World According to Pimm*

The year 2015 saw the publication of an impressive *tour d'horizon* of global ecology. Covering many areas, it assesses human impacts on biodiversity, the subject that falls within my expertise. Like all good reviews, it's well documented, comprehensive, and contains specific suggestions for future research. Much of it has a familiar feel, although it's a bit short on references from *Nature* and *Science*. But that's not what makes this review news. Rather, it's because it reached a well-defined 1.2 billion people, plus uncountable others—putting the citation statistics of other recent science stories in the shade. The publication is "On Care for Our Common Home," and its author is better known as Pope Francis.

How much ecology is there in this? And how good is it? Well, the word "ecology" (or similar) appears eighty times, "biodiversity" twelve, and "ecosystem" twenty-five. There's a 1,400-word section on the loss of biodiversity—the right length for a letter to *Nature*.

The biodiversity section starts with a statement that Earth's resources "are also being plundered because of short-sighted approaches to the economy, commerce and production." It tells us that deforestation is a major driver of species loss. It explains that a diversity of species is important as the source of food, medicines, and other uses, and that "different species contain genes which could be key resources in years ahead for meeting human needs and regulating environmental problems." A high rate of extinction raises ethical issues—in particular, the idea that our current actions limit what future generations can use or enjoy.

We learn that most of what we know about extinction comes from studying birds and mammals. In a sentence that E. O. Wilson might have written, it praises the small things that rule the world: "The good functioning of ecosystems also requires fungi, algae, worms, insects, reptiles and an innumerable variety of microorganisms. Some less numerous species, although generally unseen, nonetheless play a critical role in maintaining the equilibrium of a particular place." There is no point in a complete catalog, but this short list exemplifies its insights and comprehensiveness. Knocking pieces from any complex system—in this case, species from ecosystems—can have unexpected effects.

Technology has benefits, but Bergoglio eloquently rejects unbridled technological optimism: "We seem to think that we can substitute an irreplaceable and irretrievable beauty with something which we have created ourselves." We not only destroy habitats, but we fragment those that remain behind. The solution is to create biological corridors. He continues: "When certain species are exploited commercially, little attention is paid to studying their reproductive patterns in order to prevent their depletion and the consequent imbalance of the ecosystem."

Whereas there has been significant progress in establishing protected areas on land and in the oceans, there are concerns about the Amazon and the Congo (the last remaining large blocks of tropical forest) and about replacing native forests with tree plantations, which are so much poorer in species.

Overfishing and discarding large amounts of bycatch diminish the oceans' ability to support fisheries. Human actions physically damage the seabed across vast areas, radically altering the composition of the species living there. The section ends with a statement that might have come from a Policy Forum in *Science,* arguing as it does for increased effort and funding:

> Greater investment needs to be made in research aimed at understanding more fully the functioning of ecosystems and adequately analyzing the different variables associated with any significant modification of the environment. Because all creatures are connected, each must be [conserved], for all . . . are dependent on one another. . . . This will require undertaking a careful inventory of [species] with a view to developing programmes and strategies of protection with particular care for safeguarding species heading towards extinction.

The biodiversity section would make an outstanding course outline for my graduate course in conservation. Its coverage is impressive, its topics of global significance. Its research is strikingly up-to-date and hints at active controversies.

The encyclical includes lengthy sections on pollution, climate change, water, urbanization, social inequality and its environmental consequences, both the promise and threat of technology, intergenerational equity, policies both local and global. All these topics would appear in a course on global ecology. But this is not why its publication made news. Rather, it's an incontestable statement of the importance of science in shaping the ethical choices of our generation—for Catholics and non-Catholics alike. It asks all religions and all scientists to grasp the enormity of the problems that the science of ecology has uncovered and to seek their solutions urgently. The author deserves the last word—and it is a good one—on how we should do that:

> Nonetheless, science and religion, with their distinctive approaches to understanding reality, can enter into an intense dialogue fruitful for both. Given the complexity of the ecological crisis and its multiple causes, we need to realize that the solutions will not emerge from just one way of interpreting and transforming reality. . . . If we are truly concerned to develop an ecology capable of remedying the damage we have done, no branch of the sciences and no form of wisdom can be left out, and that includes religion and the language particular to it.

LEAKING, THINNING, SLIDING ICE

LAURENCE C. SMITH
Professor and Chair, Department of Geography; professor,
Department of Earth, Planetary, and Space Sciences, UCLA;
author, *The World in 2050*

Recently the *New York Times, Wall Street Journal, Los Angeles Times,* and other prominent news outlets around the world have been granting an abnormally high level of media coverage to scientific news about the world's great ice sheets. The news conveyed is not good.

Through unprecedented new images, field measurements, and modeling capabilities, we now know that Greenland and Antarctica, remote as they are, have already begun the process of redefining the world's coastlines. More than a billion people—and untold aspects of our economies, ecosystems, and cultural legacies—will be altered, displaced, or lost in the coming generations.

Five studies in particular commanded especial attention. One showed that the floating ice shelves ringing Antarctica (which do not affect sea level directly but do prevent billions of tons of glacier ice from sliding off the continent into the ocean) are thinning, their bulwarking ability compromised. Another, through the use of drones, satellites, and extreme field work, found pervasive blue meltwater rivers gushing across the ice surface of Greenland. A major NASA program called Oceans Melting Glaciers, or OMG, showed that the world's warming oceans—which thus far have absorbed most of the heat from rising global greenhouse-gas emissions—are now melting the big ice sheets from below, at the undersides of marine-terminating glaciers. A fourth study used historical air photographs to map the scars of 20th-century deglaciation around the edges of the Greenland ice sheet, showing that its pace of volume loss has accelerated. A fifth, a long time-horizon study, used advanced computer modeling to posit that the massive Antarctic ice sheet may disappear altogether in coming millennia, should we choose to burn all known fossil-fuel reserves.

That last scenario is extreme. But if we choose to bring it to reality, the world's oceans would rise an additional 200 feet. To put 200 feet of sea-level rise into perspective: The entire Atlantic seaboard, Florida, and the Gulf Coast would vanish from the United States, and the hills of Los Angeles and San Francisco would become scattered islands. Even 5 or 10 feet of sea-level rise would change or imperil the existence of coastal populations as we currently know them. Included among these are major cities like New York, Newark, Miami, and New Orleans in the U.S.; Mumbai and Calcutta in India; Guangzhou, Guangdong, Shanghai, Shenzen, and Tianjin in China; Tokyo, Osaka, Kobe, and Nagoya in Japan; Alexandria in Egypt; Haiphong and Ho Chi Minh City in Vietnam; Bangkok in Thailand; Dhaka in Bangladesh; Abidjan in Côte d'Ivoire, Lagos in Nigeria, and Amsterdam and Rotterdam in the Netherlands. The risk is not simply of rising water levels but also of the enhanced reach of storm surges (as illustrated by Hurricane Katrina and Superstorm Sandy); and of private capital and governments ceasing to provide insurance coverage for flood-vulnerable areas.

Viewed collectively, these studies and others like them tell us four things that are interesting and important.

The first is that ice sheets are leaky, meaning that it seems unlikely that increased surface melting from climate warming can be countered by significant retention or refreezing of water within the ice mass itself.

The second is that the pace of global sea-level rise, which has already nearly doubled over the past two decades (and is currently increasing approximately 3.2 mm/year, on average), is clearly linked to the shrinking ice volumes of ice sheets.

The third is that warming oceans represent a hitherto unappreciated feedback to sliding ice.

The fourth is that the process of ice-triggered sea-level rise is not only ongoing but accelerating. Many glaciologists now fear that earlier estimates of projected sea-level rise by the end of this century (about 1 foot if we act aggressively now to curb emissions, about 3.2 feet if we do not) may be too low.

Sea-level rise is real; it's happening now and is here to stay. Only its final magnitude remains for us to decide.

GLACIERS

ROBERT TRIVERS

Evolutionary biologist; professor of anthropology and biological sciences, Rutgers University; author, *Wild Life: Adventures of an Evolutionary Biologist*

Glaciers throughout the world are melting at an unprecedented rate. Glaciers throughout the world will continue to melt at an unprecedented rate. Try living with an average sea level 5+ meters higher.

OUR COLLECTIVE BLIND SPOT

JENNIFER JACQUET
Assistant professor of environmental studies, NYU;
author, *Is Shame Necessary?*

Scientists and the media are establishing new ways of looking at who is responsible for anthropogenic climate change. This expanded view of responsibility is some of the most important news of our time, because whomever we see as causing the problem informs whom we see as obligated to help fix it.

The earliest phases of climate responsibility focused on greenhouse-gas emissions by country and highlighted differences between developed and developing nations (a distinction that has become less marked as China and India have become two of the top three emitters). Then, in the first decade of the 21st century, the focus, at least in the U.S., narrowed to individual consumers. However, this century's second decade has brought corporate producers into the spotlight, not only for their role in greenhouse-gas emissions but also for their coordinated efforts to mislead the public about the science of climate change and prevent political action.

Although we have traditionally held producers responsible for pollutants, as in the case of hazardous waste, a debate followed about whether it was fair to shift the burden of responsibility for greenhouse-gas emissions from demand to supply. New research revealing how some fossil-fuel companies responded to climate science has placed a greater burden on the producer. Since the late 1980s, when the risks of climate change began to be clear, some corporations funded efforts to deny climate science and worked to ensure the future of fossil fuels. Producers influenced public beliefs and preferences.

One reason for the recent research into corporate influence is the growing number of disciplines (and interdisciplines) involved in climate research. While psychologists were some of the first to conduct headline-generating climate-related social science (which helps explain the focus on individual responsibility and preferences), researchers from other disciplines, like sociology and history of science, began documenting the role of corporations and a complicit media in the failure to act on climate change.

The mounting evidence for producer culpability has happened relatively quickly, but its timing remains embarrassing. Over the last two decades of the climate wars, scientists have been accused of being bad communicators, of emphasizing uncertainty, and of depressing and scaring people. I find none of these lines of argument particularly convincing. But the failure of researchers and the media, until recently, to neither see nor document industry's legerdemain as partly responsible for the stalemate over climate represents their (our) biggest failure on climate action. We might be able to blame corporate influence over politics and the media for the public-opinion divide, but that doesn't explain why researchers and journalists overlooked the role of corporations for so long. Now that we've recognized industry's important role in climate change, let's hope this doesn't regress into our collective blind spot.

THREE DE-CARBONIZING SCIENTIFIC BREAKTHROUGHS

BILL JOY
Futurist; cofounder and former chief scientist, Sun Microsystems; Greentech Venture Capital Group and Partner Emeritus, KPCB

Climate change is an enormous challenge. Rapid de-carbonization of manufacturing, electricity generation, and transportation is critical and may become a crisis because of nonlinear effects. Last year brought not-widely-disseminated news of the commercial availability of three substantial scientific breakthroughs that can significantly accelerate de-carbonization.

1. De-carbonizing Concrete; Commoditizing CO_2

After water, concrete is the most widely used material in the world. The manufacture of Portland cement for use in concrete accounts for up to 5 percent of global anthropogenic emissions. A new "Solidia cement," invented by Dr. Richard Riman of Rutgers University, can be made from the same ingredients as Portland cement and in the same kilns, but at lower temperature, while incorporating less limestone, thus emitting substantially less CO_2 in its manufacture. Unlike Portland cement, which consumes water to cure, this new cement cures by consuming CO_2. Concrete products made from it have their CO_2 footprint reduced by up to 70 percent. Thousands of tons of the new cement have been manufactured, and 2015 brought news that large manufacturers are now modifying their factories to use it instead of Portland cement to make concrete. Its widespread adoption would multiply the demand for industrial CO_2 substantially, creating a strong economic incentive for CO_2 capture and reuse.

Previous attempts at introducing radically-low-carbon cements have all failed to scale, because they required raw materials that were not ubiquitous and expensive new capital equipment, and/or because of the large range of material properties required by regulation or for specific applications. Solidia cement overcomes these problems and offers lower cost and better performance. But rapid adoption in an existing infrastructure has to be simple. In this case, only a single step of the manufacturing process—to cure with CO_2 rather than water—has to change.

Can we similarly expect to reduce the CO_2 footprint of other high-embodied-energy materials, such as steel and aluminum, while reusing the existing infrastructure? A decade-long search found no suitable candidate breakthroughs, so these de-carbonizations may unfortunately require a much slower process of redesigning products to use lower-embodied-energy materials like structural polymers and fibers.

2. Scalable Wind Turbines for Distributed Wind

More than a billion people, mostly in rural areas in the developing world, lack access to a reliable grid and electricity; it matters greatly whether they will get electricity from renewables or fossil fuels. Wind turbines today are the cheapest renewable but only in very large multi-megawatt Utility-Scale units unsuitable for distributed generation. At smaller sizes, the performance of existing wind-turbine designs degrades substantially. A new type of shrouded wind turbine, invented by Walter Presz and Michael Werle of Ogin Energy, saw its first multi-unit deployment at Mid-Scale (100kW-rated range) in 2015. This new turbine's shroud system pumps air around the turbine so that it is efficient at both Mid- and Small-Scale sizes and at lower wind speeds, thus supporting distributed generation and microgrids.

A recent analysis shows Utility-Scale Wind the cheapest renewable, with unsubsidized cost at about $80/MWh, Solar PV at about $150/MWh; conventional Mid-Scale Wind is $240/MWh—too expensive to make a substantial contribution. The new shrouded turbines provide electricity at half the cost of the conventional Mid-Scale turbines today and will be cost-competitive with Utility-Scale Wind when they are in volume production.

We need to deploy enormous amounts of renewables to fully de-carbonize electricity generation and enable the necessary decommissioning of most of the existing fossil-fuel-consuming generating equipment. Wind can be deployed extensively much more quickly, safely, and cheaply than the often proposed scale-up of nuclear energy, and can be combined with grid storage such as batteries to make it fully dispatchable. If we get serious about de-carbonization, Small- and Mid-Scale turbines can be quickly scaled to high-volume production using existing manufacturing infrastructure, much as was done for materiel during WW II. Having cost-effective wind at all scales complements Solar PV and, with grid storage, completes a portfolio that can further accelerate the marked trend toward renewables.

3. **Room Temperature Ionic Electrolyte for Solid State Batteries**

Current lithium-ion batteries use a flammable liquid electrolyte and typically incorporate materials that further increase fire danger. Most contain expensive metals, like lithium, cobalt, and nickel. Last year brought publication of the existence of a new polymer electrolyte invented by Michael Zimmerman of Ionic Materials—the first solid to have commercially practical ionic conductivity at room temperature. The polymer is also inherently safe, self-extinguishing when set on fire. It creates a chemical environment substantially different from that of a liquid, supporting novel and abundant cathode materials, like sulfur (which is high-capacity, light, and inexpensive) and new and inexpensive metal anodes, thereby supporting multivalent species, like $Zn2+$. Many desirable battery chemistries, infeasible with liquid electrolytes, are newly possible.

This scientific breakthrough, shown only in the 2030s on most battery-industry roadmaps, has long been desired because solid batteries can be much cheaper and safer and store more energy. Solid polymer batteries can be manufactured using mature and inexpensive scale-manufacturing equipment from the plastics industry.

Fifteen percent of global CO_2 fossil-fuel emissions come from wheeled transportation. India and China's fleets will grow substantially in the years ahead; whether energy for these additional vehicles is provided by renewables or fossil fuels will make a significant difference in global emissions. Low-cost, safe, and high-capacity batteries can greatly accelerate the electrification of transportation and these fleets beyond the current modest projections.

In the 21st century, we need to stop combusting fossil fuels. Electrochemistry—both better batteries and fuel cells—has far greater potential than is generally realized and can displace most combustion.

There are other gas- and liquid-based technologies we can hope to convert to solid-state to reduce their CO_2 impact—such as cooling, which generally uses a liquid-gas phase transition today. I hope the future brings news of a solid-state cooling breakthrough that, like the above technologies, can be quickly taken to scale.

JUICE

JAMES CROAK
Artist

In one hand you're holding a gallon of gasoline weighing 6 pounds, in the other a 3-pound battery; now imagine them containing equal energy. Spoiler alert: They already can. The most exciting and far-reaching scientific advance is the dramatic increase in electric-battery density, allowing it to displace gasoline and solving the problems of night electricity, vehicle range, and becalmed windmills.

Electric-car range increases about 9 percent per year and has reached a point where one can imagine round-trips that don't involve a flatbed. But the public was startled in 2011 when a seven-figure prize was claimed from Green Flight Challenge, which had offered it to an aircraft that could fly 200 miles in under two hours with a passenger, using less than a gallon of fuel. Three planes competed—two electric and one hybrid—with only the electric planes finishing within the allotted time. The winner averaged 114 mph on a plug-in electric plane *sans* a gas engine. This was a Tom Swift fantasy five years earlier, because the weight of the batteries was too much for the plane—even if they had been able to be crammed into fuselage. Their weight and size shrank, while the energy storage increased.

Battery density now peaks at about 250 watt-hours per kilogram, up markedly from 150 wh/kg a few years ago but still far below petroleum, which is 12,000 wh/kg. One company is about to release a 400 wh/kg, but batteries under development could pass the energy density of fossil fuels within a few years.

The most exciting and counterintuitive battery invented is the lithium-air, which inhales air for the oxygen needed for its chemical reaction and exhales the air when finished. This should ring a bell: Gas engines inhale air and add a gas mist; the expanding air creates power but then expels an atmospheric sewer. The lithium-air battery is solid-state and exhales clean air. MIT has already demonstrated a lithium-air battery with densities of over 10,000 wh/kg.

Batteries need not have the energy density of gasoline in order to replace it. The physics of gasoline power are lame; only 15 percent of the energy in your tank powers the car down the highway; the rest is lost to heat, engine and transmission weight, friction, and idling. As a practical matter, batteries in the labs are already beyond usable energy density of fossil fuels, an energy density that results in a 500-mile range for an electric car with a modest battery and probably more for a small plane.

The second substantial change is that increased battery density has lowered both the size and cost of electrical storage, creating the bridge between intermittent wind and daytime photovoltaic energy, and the round-the-clock current demands of the consumer.

Windmills produce prodigious electricity during a good blow but bupkis when becalmed; the batteries provide steady current until a breeze appears. A new battery installation at the Elkins, West Virginia, windfarm keeps the 98-megawatt turbines as a constant part of the overall grid supply, with pollutant-free electricity and the reliability of a conventional fossil-fuel plant.

Also, fossil-fuel plants run at higher capacity than needed, in case of a spike in demand. A new megawatt battery installation in the Atacama Desert of Chile brought stability to the grid and a reduction in fuel usage.

The hoped-for green revolution is suddenly here, improbably due to the humble battery. A century ago, there were more electric cars on the road than gasoline cars. Very soon, we will be back to the future.

A CALL TO ACTION

HANS ULRICH OBRIST

Curator, Serpentine Gallery, London; author, *Ways of Curating*

The publication in 2015 of a paper by Mark Williams *et al.* titled "The Anthropocene Biosphere"* provides more evidence that the changes wrought upon the climate by human civilization are set to produce a sixth mass extinction. According to one of the paper's co-authors, geologist Peter Haff, we have already entered a period of fundamental changes that may continue to alter the world beyond our imagination. All of us can provide anecdotal evidence of the shifts in our environment. In December I received a call from a friend in Engadin, Switzerland, where Nietzsche wrote *Thus Spoke Zarathustra;* at an altitude of 2,000 meters, there was no snow. Meanwhile, in Hyde Park, the daffodils were blooming.

As the artist, environmentalist, and political activist Gustav Metzger has been saying for years, it is no longer enough just to talk about ecology: We need to create calls to action. We must consider the potential for individual and collective agency to effect changes in our behavior and develop adaptive strategies for the Anthropocene age. To quote Metzger, we need "to take a stand against the ongoing erasure of species, even where there is little chance of ultimate success. It is our privilege and our duty to be at the forefront of the struggle." We must fight against the disappearances of species, languages, entire cultures; we must

---

* http://anr.sagepub.com/content/2/3/196

battle the homogenization of our world. We must understand this news as part of a broader continuum. The French historian Fernand Braudel advocated the *longue durée,* a view of history which relegates the historical importance of "news events" to a place beneath the grand underlying structures of human civilization. Extinction is a phenomenon belonging to the *longue durée* of the Anthropocene, the symptoms of which we are beginning to experience as news. By connecting the news to the *longue durée,* we can formulate strategies to transform our future and avert the most catastrophic extinction scenarios. By understanding the news, we can act upon it.

Art is one means by which we reimagine existing paradigms to accommodate new discoveries, the thread connecting the now to the past and future, the thread linking news events to the *longue durée.* Art is also a means of pooling knowledge, and it is, like literature, news that *stays* news. When Shelley stated that "poets are the unacknowledged legislators of the world," he meant something like this: that writers and artists reimagine news in ways that change how we perceive the world, how we think and act.

Among my great inspirations is Félix Fénéon, a fin-de-siècle French editor (and the first publisher of James Joyce in France), art critic (he discovered and popularized the work of Georges Seurat), and anarchist (put on trial, he escaped prosecution after famously directing a series of barbs at the prosecutor and judge, to the jury's great entertainment). Fénéon was a master of transformation. He transformed the news into world literature via his series of prose poems. In 1906 he was the anonymous author of a series of three-line news items published in *Le Matin* which have since become famous. Those brief reports adapted stories of contemporary murder and misery into prose poems that will last forever. Lawrence Durrell's *Alexandria Quartet* transformed the Copernican breakthroughs of Einstein and Freud into fiction. By translating events that are ephemeral and local in their initial impact into that which is universal and enduring, we can make news into culture.

John Dos Passos gave lasting form to events that seemed characterized by a fleeting immediacy. In his *U.S.A.* trilogy, Dos Passos pioneered new styles of writing that sought to capture the experience of living in a society overwhelmed by the proliferation of print media, television, and advertising. In his "newsreel" sections, the author collages newspaper clippings and popular song lyrics; elsewhere he pursues his experiments in what he called the "camera eye," a stream-of-consciousness technique that attempts to replicate the unfiltered receptivity of the camera—which makes no distinction between what is important and what is not. Later this material is transformed into stories. According to the filmmaker Adam Curtis, the *U.S.A.* trilogy identifies

> the great dialectic of our time, which is between individual experience and how those fragments get turned into stories, . . . [W]hen you live through an experience you have no idea what it means. It's only later, when you go home, that you reassemble those fragments into a story. And that's what individuals do, and it's what societies do. It's what the great novelists of the 19th century, like Tolstoy, wrote about. They wrote about that tension between how an individual tells the story of an event themselves, out of fragments, and how society then does it.

The Lebanese-American poet, painter, novelist, urbanist, architect, and activist Etel Adnan speaks about the process of transformation as the "beautiful combination of a substratum that is permanent and something that changes on top. There is a notion of continuity in transformation." In her telling, transformation describes the relationship between the *longue durée* of history, current news events, and action that can transform the future. She shows us how dialogue can produce new strategies that can preserve difference and help act against extinction, while also acknowledging that change is inevitable. If we are to develop radical new strategies to address one of the most important issues of our time, then we must go beyond the fear of pooling knowledge among disciplines. If we do not pool knowledge, then the news is just news: Each new year will bring reports of another dead language, another species lost. While writing this text, I received an email in the form of a poem from Etel Adnan which expresses this beautifully:

WHERE DO THE NEWS GO?

> News go where angels go
> News go into the waste-baskets of foreign
> embassies
> News go in the cosmic garbage that the
> universe has become
> News go (unfortunately) into our heads.

A BRIDGE BETWEEN THE 21ST AND 22ND CENTURY

KOO JEONG-A
Conceptual artist

Aristotle discussed magnetism with Thales of Miletus. Oriental medicine refers to the meridian circles and was treating by using the magnetic field before the invention of the acupuncture needle in the Iron Age. As the Italian philosopher Benedetto Croce wrote, "All history is contemporary history." The magnet's cryptographic character—relevant in the computer network, medical devices, and space expeditions through electromagnetic fields that link multiple cultural devices in our saturated era—as a decorous bridge between the 21st and the 22nd century, will still innovate. Far from extreme division, magnet-espoused technology would make peace in our world.

THE GREATEST ENVIRONMENTAL DISASTER

RICHARD MULLER

Physicist, UC Berkeley; author, *Energy for Future Presidents*

The news stories from China are horrific. The best estimate is that *on average*, 4,400 people die every day from air pollution in that country. That's 1.6 million per year. Every time I hear of some tragedy that makes headlines, such as a landslide in Shenzhen that killed 200 people, I think to myself, "Yes—and today 4,400 people died of air pollution and it didn't make the news."

This is not the old eye-burning, throat-irritating air pollution of yesterday. Today's pollutant is known as PM2.5—particulate matter 2.5 microns and smaller. It is produced by automobiles, by construction, by farm work, but the greatest contributor by far is coal, burned by industry and for electric-power production. PM2.5 wasn't even listed as a major pollutant by the U.S. Environmental Protection Agency until 1997. It was present but just not fully proved to be as deadly as it is.

We now know that on a bad day in Beijing, such pollution hurts people as much as smoking two packs of cigarettes a day. Bad air triggers strokes, heart attacks, asthma, and lung cancer. Look at the causes of death in China and you'll see a remarkable excess of such deaths, despite the fact that obesity is uncommon compared with that in the U.S.

We know about the health effects from some remarkable studies. In the U.S., we saw decreases in health problems when factories and coal plants were temporarily shut down (the 1993

"Six Cities Study"). In China, we have the Huai River Study, in which the Chinese policy of giving free coal to households north of the Huai River, but none to the south, resulted in a reduced average lifetime in the north of 5.5 years.

Also remarkable is China's openness with their air-pollution data. Every hour, they post online more than 1,500 measurements of PM2.5 (as well PM10, SO_2, NO_2, and ozone) all across their country. China may be a closed society in many ways, but they seem to be crying out for help. At Berkeley Earth, we have been downloading all these numbers for the past year and a half, and the patterns of severe pollution are now clear. It is not confined to cities or basins but widespread and virtually inescapable. Ninety-seven percent of China's population breathes what the EPA deems as "unhealthy air" on average. In contrast, the democracy of India reports few PM2.5 measurements. I suspect they have them but are simply not making them public. They do publish results for Delhi, and virtually every time I look, the pollution there is worse than it is in Beijing.

People suggest a switch from coal to solar, but this is too expensive for China to afford. In 2015, solar power contributed less than 0.2 percent to their energy use, and solar plants are going bankrupt as Chinese subsidies are withdrawn. Wind power is expanding, but wind's intermittency is a big problem, and the use of energy storage drives up cost. Hydro is hardly an environmental choice; the Three Gorges Dam displaced 1.2 million people (voluntarily, the Chinese tell us) and destroyed 13 cities, 140 towns, and 1,350 villages. Their new Mekong River dam is expected to wreak havoc throughout Myanmar, Thailand, and Vietnam.

The best hopes consist of natural gas, which China has in abundance, and nuclear power, which is under rapid development. PM2.5 from natural gas is reduced by 1/400 compared with coal—and it reduces greenhouse emissions by a factor of 2 to 3. China is desperately attempting to extract its shale gas but is doing miserably; the only true master of that technology is the U.S., where it has triggered an enormous and unexpected drop in the price of both natural gas and oil. Nuclear power, once despised by environmentalists, is gaining traction in the U.S., with many past opponents recognizing that it offers a way to reduce carbon emissions significantly. China is surging ahead in nuclear, with thirty-two new plants planned. Although such plants are reported to be expensive, the Chinese know that the high cost is only in the capital cost—that amortized over twenty-five years, nuclear is as cheap as coal, and much cheaper when you add in the environmental costs.

Air pollution will be a growing story. China also has plans, on paper, to double its coal use in the next fifteen years. They will cancel those if they can, but they also worry that slower economic growth could threaten their form of government. As bad as the pollution has been so far, I worry that we ain't seen nothing yet.

The United States is sharing its nuclear technology, and I expect that in two decades China will be the principal manufacturer of nuclear power plants around the world. But we need to set a better example; we need to show the world that we consider nuclear to be safe. And we need to share our shale-gas technology far more extensively. Too often we read the pollution headlines, shake our heads, perhaps feel a little *schadenfreude* toward our greatest economic adversary, and then we forget about it.

Someday global warming may become the primary threat. But it is air pollution that is killing people now. Air pollution is the greatest environmental disaster in the world today.

TECHNOBIOPHILIC CITIES

SCOTT SAMPSON

Vice president of research and collections, Denver Museum of
Nature and Science; dinosaur paleontologist; science communicator;
author, *How to Raise a Wild Child*

The news cycle regularly features stories about reinventing 21st-
century cities. Among societal issues, perhaps only education is
targeted more frequently for reform. And for good reason.

Since 2008, more people have lived in cities than not. By
the end of this century, cities will generate nearly 90 percent
of population growth and 60 percent of energy consumption.
While these bustling hubs of humanity function as the planet's
innovation centers, they're also responsible for the lion's share of
environmental damage. By some estimates, today's cities generate around 75 percent of global carbon-dioxide emissions, along
with countless other pollutants. They consume vast expanses
of forests, farmland, and other landscapes, while fouling rivers,
oceans, and soils. In short, if we don't get cities right, it's hard to
imagine a healthy future for humanity, let alone the biosphere.

By my reading, most of the press surrounding the reinven-
tion of cities can be grouped into two camps. One camp calls for
"smart," "digital," and "high-tech" cities. Here the emphasis is on
information and communication technologies with the potential
to boost urban functioning. Fueled by the recent tsunami of civic
data (climate information, traffic patterns, pollution levels, power
consumption, etc.), key areas cited for high-tech interventions include flows of people, energy, food, water, and waste. Advocates

imagine cities that can talk, providing live status updates for pollution, parking, traffic, water, power, and light. Thanks to such innovations as ultra-low power sensors and Web-based wireless networks, smart cities are rapidly becoming reality.

From the other camp we hear about the need for "green," "biophilic," even "wild" cities, where nature is conserved, restored, celebrated. Of course, cities have traditionally been places where the wild things aren't, engineered to wall humans off from the natural world. Yet recent and rapidly accumulating research documents the positive health effects of regular contact with urban nature. Benefits include reduced stress levels, stronger immune systems, and enhanced learning. Perhaps most important are the myriad physical, mental, and emotional benefits that seem essential to a healthy childhood. Proponents of the green-city camp also argue that many of the pressing issues of our time, among them climate change, species extinctions, and habitat loss will not—indeed, cannot—be addressed unless people understand and care about nearby nature.

So there you have it. Big Data versus Mother Nature. Two views on the future of cities, apparently at opposite ends of the spectrum. One values technological innovation, the other biological wisdom and nature connection. Yet on close inspection these perspectives are far from mutually exclusive. In fact, they're complementary.

It's entirely possible for cities to be both high-tech and nature-rich. Today, few proponents of green cities claim we need to go "back to nature." Rather they argue for going forward into a future rich in both technology and nature. New terms like "technobiophilic cities" and "nature-smart cities" are emerging to describe this blended concept, urban settings where the natural and digital are embraced simultaneously.

Yes, nature-smart cities will have plenty of green roofs, green walls, and interconnected green spaces. Seeding native plants attracts native insects, which in turn entices native birds and other animals, transforming backyards, schoolyards, and courtyards into miniature ecosystems. These nuggets of urban nature, in addition to improving the health of humans, are the last good hope for scores of threatened species. In addition, cities rich in nature can leverage smart technologies to help urbanites switch to renewable energy sources—wind, sun, water, and geothermal. Green transportation reduces carbon emissions and improves the environment. Green buildings can act like trees, running on sunlight and recycling wastes, so that cities function like forests.

Interestingly, both views of our collective urban future highlight the importance of an informed and engaged citizenry. Digital technologies and Big Data may well put control back into the hands of individuals—for example, through greater participation in local governance ("E-Governance"). Similarly, citizen scientists and citizen naturalists can play important roles in restoring plants and animals, monitoring these species, and making adjustments to improve the quality and quantity of nearby nature. Here, then, is a potent pathway to help people act on the basis of robust scientific data (and boost science literacy along the way).

In short, there's much more than hot air in all the news about reimagining the future of cities. At least within urban settings, Mother Nature and Big Data have the potential to make excellent bedfellows. Indeed our survival, and that of much of Earth's biodiversity, may depend on consummating this union. If successful, we'll witness the birth of a new kind of city, one in which people and nature both thrive.

LENR COULD SUPPLANT FOSSIL FUELS

CARL PAGE

President, Anthropocene Institute; engineer; entrepreneur; cofounder, eGroups

Climate collapse demands a supply of energy far cheaper than fossil fuels, resistant to bad weather and natural disaster, and sustainable in fuel inputs and pollution outputs. Can a new, poorly understood technology from a stigmatized field fulfill the need? The Low Energy Nuclear Reaction (LENR) could help at large scale very quickly.

In 1989, Stanley Pons and Martin Fleischmann provided an initial glimpse of an unexpected reaction dubbed cold fusion, which makes lots of heat and very little radiation. LENR is being pursued quietly by many large aerospace companies, leading automakers, startup companies, and, to a lesser extent, national labs. Over the years, many teams have observed the reaction by various means, and a consistent pattern has emerged. Experiments have become more repeatable, more diverse, more unambiguous, and higher in energy. There are no expensive or toxic materials or processing steps, so it could be the move beyond fossil fuels we have been waiting for. No government-regulated materials are used, so a quick path to commercialization is possible.

Familiarity with hot fusion led to initial false expectations. Early hasty replication work at MIT was declared a failure when heat but no high-energy neutrons were detected. The reaction requirements were not known at first, and many attempts failed to reach fuel-loading and ignition-energy requirements. Even when the basic requirements were met, nanoscale features varied in materials and made the reaction hard to reproduce. Pons and Fleischmann had trouble repeating their own excess-energy results after they used up their initial lucky batch of palladium. Today we understand better how material defects create required high-energy levels.

In many experiments with LENR, observed excess heat markedly exceeds known or feasible chemical reactions. Experiments have gone from milliwatts to hundreds of watts. Ash products have been identified and quantitatively compared to energy output. High-energy radiation has been observed and is entirely different from hot fusion.

Michael McKubre at SRI International teased out the required conditions from the historical data. To bring forth LENR reactions that produce over-unity energy, a metal lattice is heavily loaded with hydrogen isotopes. It is driven far out of equilibrium by some excitation system. High proton flux and electromigration of lattice atoms are also found in successful demonstrations.

Dr. Melvin Miles quantitatively characterized the outputs of LENR in meticulous 1995 experiments at China Lake. LENR releases helium-4 and heat in the same proportion as hot fusion, but neutron emissions and gamma rays are at least 6 orders of magnitude less than expected.

Successful excitation systems have included heat, pressure, dual lasers, high currents, or overlapping shock waves. In order to create conditions that drive the reaction, materials have been treated to create energy-concentrating flaws, holes, defects, cracks, and impurities and to increase surface area. Providing a high flux of protons and electron current is also characteristic of successful demonstrations. Solid transition metals, including nickel and palladium, host the reaction. Ash includes ample evidence of metal isotopes in the reactor which have gained mass, as if from neutron accumulation—as well as enhanced deuterium and tritium. Tritium is observed in varying concentrations. Weak X-rays are observed, along with tracks from other nuclear particles.

LENR looks like fusion—judging, as a chemist might, by the input hydrogen and output helium-4 and transmutation products. It looks not at all like fusion when judging it as a plasma physicist might—by tell-tale radioactive signatures. Converting hydrogen to helium will release lots of energy no matter how it's done. LENR is not zero-point energy or perpetual motion. The question is whether that energy can be released with affordable tools.

Plasma physicists understand hot thermonuclear fusion in great detail. Plasma interactions involve few moving parts, and the environment is random, so its effect is zeroed out. In contrast, modeling the LENR mechanism will involve solid-state quantum mechanics in a system of a million parts being driven far out of equilibrium. In LENR, a nanoscale particle accelerator can't be left out of the model. A theory for LENR will rely on intellectual tools that illuminate X-ray lasers or high-temperature superconductors or semiconductors.

Many things need to be cleared up. How is the energy level concentrated enough to initiate a nuclear reaction? What is the mechanism? How do output energies in the MeV range produce heat without obvious high-energy particles? Peter Hagelstein at MIT has been working hard at a "Lossy Spin Boson Model" for many years to cover some of these gaps.

Robert Godes at Brillouin Energy offers a theory that

matches observations and suggests an implementation: the controlled electron-capture reaction. Protons in a metal matrix are trapped to a fraction of an angstrom under heat and pressure. A proton can capture an electron and become an ultracold neutron that remains stationary but without the charge. That allows another proton to tunnel in and join it, creating heavier hydrogen and heat. Neutron accumulation creates in succession deuterium, tritium, and hydrogen-4. Hydrogen-4 is new to science and is predicted (and observed?) to beta-decay to helium-4 in about 30 milliseconds. All this yielding about 27 MeV in total per atom of helium-4, as heat.

The proton-electron capture reaction is common in the Sun, and predicted by supercomputer simulation at Pacific Northwest National Laboratory [PNNL]. It is the reverse of free-neutron beta decay. Such a reaction is highly endothermic, absorbing 780 KeV from the immediate surroundings.

Fission experts expect hot neutrons to break up fissile atoms. LENR does it backward—ultracold neutrons (which cannot be detected by neutron detectors but can readily be confirmed by isotope changes) are targets for hydrogen. Hence helium is produced with the tools of chemistry and without overcoming the Coulomb positive-particle repulsion force. And without requiring or producing radioactive elements.

The theory of LENR's exact mechanism is still in dispute; no theory pleases everyone. Output power levels are usually below commercial viability, and many different methods produce the characteristic excess heat. Just as gold was mined before geology was understood but got a lot more predictable afterward, LENR methods are reinforced by occasional success. The complete explanation of the mechanism will happen when a serious effort is made to prove (or disprove) the candidate theories. Nowadays we are forced to rely on entrepreneurial zeal instead of orderly science, because of the hyperconservatism of science-funding agencies. So there is no coordinated effort to efficiently focus on finding the correct theory. Collaboration helps, but that is one thing secretive companies are bad at.

It is strange that LENR is neglected by the DOE, industry, and the Pentagon. But no stranger than the history of nuclear power. If it weren't for the leadership of Admiral Rickover and his personal friends in Congress, nuclear-fission power for submarines and power plants would never have seen the light of day. Nevertheless, progress will be made by private enterprise in lieu of government support. Sadly, that means you cannot stay up-to-date by relying on a subscription to *Science*. But stay tuned.

EMOTIONS INFLUENCE ENVIRONMENTAL WELL-BEING

JUNE GRUBER

Assistant professor of psychology, University of Colorado, Boulder

We know that emotions can influence individual well-being. Across numerous studies, we see that the intensity and flexibility of our emotions have robust effects on a wide range of cross-sectional and longitudinal well-being outcomes. Furthermore, an optimal diversity of (positive and negative) emotional experiences in everyday life promotes greater subjective well-being and decreased psychopathology symptoms. But are the effects of emotion on well-being specific to individual-level outcomes?

Recent scientific news suggests the answer is No: Emotions also influence environmental well-being outcomes. Psychological processes, including our emotional states, play an important and previously understudied role in our response to pressing environmental issues. For example, exposure to scenes of environmental destruction engages distinct neural regions (e.g., anterior insula) associated with anticipating negative emotions: This, in turn, predicts individuals' donations to protect national parks. Thus, negative rather than positive emotional responses may drive pro-environmental behaviors (as suggested by powerful work conducted by Brian Knutson and Nik Sawe). Such findings come on the heels of task-force reports underscoring the need for an affective level of analysis, given the collective impact of emotion-relevant processes (such as emotion regulation and responding) on shaping broad-based environmental outcomes.

Important, too, are advances in psychology that recommend applying insights about individual affective reactions to spur public engagement in pro-environmental behaviors.

This burgeoning work at the intersection of affective science and environmental psychology shows that emotions can improve environmental health by shaping our emotional reactions toward environmental issues as well as the frequency and degree of conservationist behaviors. Yet much work remains to be done, including mapping the reciprocal relationship between our emotions and environmental choices in decision making and policy planning. We also need to learn more about how rapid changes in immediate environmental surroundings (e.g., access to clean water, local air pollution) might have reciprocal downstream effects on affective states and motivated behaviors. And we must further investigate whether and how individual judgments and real-world choice behaviors can scale to aggregate policy level.

As environmental concerns grow—rapid deforestation, increasing carbon-dioxide emissions, habitat destruction, threats to critical areas of biodiversity—insights from affective science will become more and more important. The time has come for social scientists to join the ranks of engineers, natural scientists, and policymakers seeking to preserve and enhance environmental well-being.

GLOBAL WARMING REDUX: A SERIOUS CHALLENGE TO OUR SPECIES

MILFORD H. WOLPOFF

Paleoanthropologist; professor of anthropology, adjunct associate
research scientist, Museum of Anthropology, University of Michigan;
author, *Race and Human Evolution*

The human species has successfully dealt with twenty or more
distinct episodes of global warming, but in circumstances that
no longer exist.

There is no real difficulty in identifying the most important
news of 2015. Global warming is the news that will remain
news for the foreseeable future, because our world will continue
to warm at a rate never before seen, without (at least at the
moment) a foreseeable end. Paleoanthropology is a comparative
science, and comparing most past episodes of global warming
to the global warming in today's news leads me to question
whether (and how) we may survive this one. And my doubts
are not solely because of a point the news also recognizes: The
rate of temperature change is much faster than humans have
ever experienced.

Prior episodes of significant global warming within the
Pleistocene (more-or-less the last 2 million years) have invariably followed cooling periods with glacial advances. During the
Pleistocene, the human lineage successfully adapted to changing
environments (including climate) and evolved to take advantage
of the opportunities afforded by the changes, even as *Homo* pop-

ulations reacted to the constraints those changes created, evolving diverse adaptations to the different climates and ecological circumstances they encountered. Improvements in communication skills and planning depth, retention of deep history with tales, poems, and song, and other aspects of cultural behavior dispersed throughout humanity.

At any particular time, the Pleistocene world population was quite small: an estimated 1 to 2 million during all but the most recent episodes of cooling and warming cycles, half or more living in Africa. Scarce on the ground, with little ecological impact, and with vast habitable areas unoccupied by human groups, the human reactions to periods of global warming often were simply population migration. These had important consequences for human evolution, because our particular brand of evolutionary change—the unique human evolutionary pattern—began with the initial and ongoing geographic dispersals of human populations. In many mammalian species, significant range expansion such as the human one resulted in geographic isolation for many groups and the formation of subspecies and ultimately of species. In humans, these processes were mitigated by continuing population interconnections created by gene flow, in some cases the result of population movements and in others because expanding human populations grew to encounter each other. The unique human evolutionary pattern was created as adaptive genes and behaviors, under selection, spread throughout the human range. Genetic changes adaptive for the entire human species were able to disperse throughout it, no matter how varied individual populations may have become, because population contacts allowed it and natural selection promoted it.

But this long-lasting pattern has been disrupted as humans gained control of their food resources and began the accelerated increase in numbers so evident today. This is recent—so recent that a rapidly growing humanity has yet to encounter significant climate change, until now. Today's headlines make it clear that the change we are encountering is global warming. It is not at all evident that the adaptive successes of the past will guarantee a successful reaction to the changes coming upon us today. The world is quite different. Adaptive strategies that once underlay a successful strategy for the human species may no longer be possible; a vastly larger number of humans probably precludes similar success from the same strategies, even absent the rapid rate of climate change we are encountering.

The fact is that the present is not a simple extension of the past. Conditions are radically different, and the strategies that promoted a successful balance of population variation within the human species and a successful adaptation for all populations throughout the Pleistocene may, today, create competition between human populations at a level that could make the lives of the survivors quite unpleasant.

Of course, nothing like this is inevitable, or even necessarily probable, but its possibility looms large enough to be taken seriously. We need to learn from the past without trying to repeat it.

BLUE MARBLE 2.0

GIULIO BOCCALETTI

Physicist, atmospheric and oceanic scientist; global managing director for water, The Nature Conservancy

The Blue Marble was the first full photograph of the Earth from space. The Apollo 17 mission took it on December 7, 1972. It was not the first photograph of the planet; by then, the first image of Earth as seen from the Moon had been widely circulated. In 1969, the lunar-landing astronauts took the famous shot of "Earthrise," capturing the solitary fragility of our planet as it rises from darkness. But the Blue Marble was a photograph of a different sort—comprehensive in scope yet detailed in nature, giving it an unusually high density of information and a powerful evocative quality. It symbolized the beginning of the Anthropocene: During the 1970s, humankind began recognizing its role in the planetary ecosystem and wrestling with the question of its impact on a finite and vulnerable planet.

The Blue Marble shows the planet from the Mediterranean to Antarctica, with the African continent and the Arabian Peninsula in the foreground and the Indian subcontinent and the Southern Ocean as frames. It provides an integrated single view of the planet's atmosphere in its spellbinding complexity: the intertropical convergence zone, where moist air flowing equatorward from north and south rises in a narrow band of convective plumes that give the characteristic thunderous rainy weather to the tropics; the Sahara and Kalahari deserts some 30° north and south of the equator, where that same air subsides from its

49

poleward flow, drying out any remaining moisture as it completes the cycle of the Hadley cell, the atmospheric overturning circulation spanning the tropics. A tropical cyclone fed by the warm surface waters of the Arabian Sea is visible in the top right quadrant. There are the mid-latitude weather systems of the roaring forties over the Southern Ocean, marked by visible fronts, altocumulus, and cirrocumulus clouds. The contour of Antarctica is revealed in full view of the Sun. It is a compendium of Earth's climate in a single shot.

Iconic geographic images can reframe how we conceive of our place on the planet. They are a recurring cultural phenomenon and a moment of synthesis revealing the preoccupations of those who produced them. While we cannot know how widespread its adoption was, the first known map of the world, the 6th-century B.C.E. "Imago Mundi" from Mesopotamia, shows the city of Babylon as it relates to the surrounding cities, the Euphrates, and the Persian Gulf, synthesizing in one image the primary elements of threat and survival of an entire civilization. The Peutinger Map, almost ten centuries later, revealed the extent of the Roman Empire in one image, providing a map organized around the great land routes that represented the strength of Roman logistics and connected the empire.

A different type of picture made headlines around the world in 2015—one that is equally representative of our preoccupation with our sustainability, and evocative of the challenges ahead.

"Tom and Jerry" went up in space on March 17, 2002. Two identical satellites formed the basis for the Gravity Recovery and Climate Experiment—GRACE, for short. The satellites orbit the Earth sixteen times a day at an altitude of more than 500 km, sending back a map of the distribution of mass on the planet due to the variation in distribution of rocks and water. They produce this data by measuring distortions in the gravitational field caused by slight differences in the distribution. As the two satellites pass over these differences, the first is slightly accelerated or decelerated with regard to the second. By measuring their relative distance to an astonishing level of accuracy—the satellites can detect a micron difference over 200 km—they provide an integrated, point-wise measurement of the gravitational field of the planet, a planetary CT scan of sorts. It is one of the great successes of modern geodesy that the resulting measurement can be inverted, filtered, and analyzed to reveal the complex three-dimensional structure of the Earth.

One of the crucial applications of this technology has been to diagnose groundwater storage in the great aquifers of the world. Water is of course different in density from the surrounding rock, thus leading to slight effects on the gravitational field, which GRACE can detect. Global hydrological assessments have always been hampered by the complex and local nature of the resource, making syntheses of the state of the world difficult to construct and of limited utility. When they do exist, they tend to be encyclopedic in nature, often published in ponderous volumes, listing individual rivers and nations as chapters, and based on single numbers, painstakingly inferred, compiled, or estimated from sparse measurements and models—hardly an evocative synthesis to stimulate a humanist debate on our place on the planet.

But in the almost decade and half since its launch, the long data sets offered by GRACE have provided the first integrated image of the state of groundwater use. It is an image global in scope yet local in nature—a detailed, realtime diagnostic of the planet. And GRACE has shed light on the most obscure part of the water cycle—that hidden underneath the Earth's surface—

and this is the picture that has made the news. It shows that one in three large aquifers in the world appear to be stressed, depleted by people drawing water for human use. California's Central Valley, the Arabian Peninsula, and the Indus basin share a common fate, uniting vastly different economies and societies in a planetary challenge: our persistent inability to manage finite resources.

While scientific practice will be integrated rather than dramatically changed by GRACE's data—remote sensing still requires significant processing and integration with land-based measurements to be operationally useful—the resulting images have already started to change the narrative on sustainability. So these data will continue to be news for the coming years. Like the ancient maps and the Blue Marble, they provide a powerful explanatory visual framework for an existential concern of our time: that another finiteness of our planet consists of the water resources we all share. GRACE has shown us that, indeed, we live on a fragile blue marble—one that is drying at an alarming rate.

HIGH-TECH STONE AGE

TOR NØRRETRANDERS

Science writer; author, *The Generous Man: How Helping Others Is the Sexiest Thing You Can Do*

The real news is old news: We belong here on this planet; we are natives. The recent news is that we are finding ways to behave as natives by using new technologies to live in an old way, a High-Tech Stone Age. Basically, it is about returning to our old niche of energy, matter, and information by using brand new technologies. Illustrative examples are food, light, and relationships.

1. **Food:**

   We used to live as hunter-gatherers, foraging for a rich variety of wild plants and animals. Now, through agriculture, we have become dependent on a select few domesticated plants and animals (more than half the calories we eat come from four crops). The machinery and fossil-fuel use involved in running nature according to our will is rising steeply, soils are eroding, and monoculture allows for pests.

   Returning to a foraging lifestyle will be difficult with 7 billion people on the planet. But a wide variety of technologies, from the simple to the complicated, offers new possibilities: Leading chefs rediscover forgotten resources in the wild—for instance, edible insects and little-known marine animals. Information technology makes foraging easier. Urban agriculture is on the rise. Many people turn away from the kinds of food that arose with agriculture (the starch in bread, pasta, rice, corn, and potatoes). Thus the old niche of wild foods and perennial plants is becoming relevant again, through the crafts of chefs and scientifically based techniques like fermentation. In the long run, unregulated growth of biomass (as opposed to the highly structured monocultures) may provide a higher yield of edible biomass. The Stone Age strategy is to let nature grow as it will. The high-tech hack is to post-process the available biomass into edibility for humans (select, cook, ferment, break down with enzymes, etc.).

2. **Light:**
    Natural light from the Sun, bonfires, and candles is thermal radiation, exhibiting a continuous spectrum. Look at it through a prism and you see a rainbow. The incandescent lightbulb is the same, since it is also thermal. But energy-saving lightbulbs and other fluorescent lights do not provide light with a continuous spectrum; they give a line spectrum, with only some of the colors of a rainbow. Thus there has been a loss of light quality and color-rendering ability in modern lighting. The incandescent bulbs have been phased out, but the replacement (energy-saving lightbulbs) gives bad light.
    LEDs have the potential to solve the problem by producing light with almost continuous spectra and with a low energy use. However, present LED light

for home and office use is not yet of a high enough quality in terms of color rendering. Our perception has adapted to seeing objects in the light from sources giving out a full rainbow, but LEDs are not there yet. They will get there, and the next wave of lighting technology will be better at producing a continuous spectrum. The use of quantum dots—artificial atoms—will allow the production of light that looks thermal but without the same energy waste as thermal sources. Solid-state lighting, like LED and quantum dots, can re-create the kind of light we have adapted to as hunter-gatherers, but with a small use of artificial energy.

3. Relationships:

The flat, peer-to-peer-based network of relationships found in hunter-gatherer cultures is ideal for regulating hunting and gathering. But the advent of agriculture meant centralization, with cities, depots, kings, and control. Thus, social structure lost the dependence on bottom-up self-organization and became reliant on top-down, rule-based societies. They are good at many things, but not for keeping civil society vibrant and alive. Also, regulating common resources is sometimes difficult for the anonymous state and market. A growing emphasis on communities that govern commons—sometimes called commonities—is a result of the climate challenge. Headquarters have been disappointing in their ability to take real action, but windmills, city gardens, and the sharing economy are no longer just naïve and vain attempts; they are changing social structure. With the advent of decentralized production (3D printers, fermentation hubs, Web-based culture), the traditional globalization trend will yield to a localization trend.

Information technology will allow humans to return to a niche of decentralized, self-organized production adjusted to the local environment. To close the loop of matter flows, local regulation is essential. The real news is that new technologies and new social strategies allow us to return to a very old resource base: the decentralized solar energy and the local flow of matter and information.

It is good news.

THE DEMATERIALIZATION OF CONSUMPTION

RORY SUTHERLAND

Executive creative director, OgilvyOne, London; vice-chairman, Ogilvy Group, U.K.; columnist, *The Spectator*

Sometime early in this century, it seems, the U.K. may have reached "peak stuff." It is a complex calculation, of course, but it seems that although the world's oldest industrialized economy had grown throughout most of that period, its consumption of raw materials and fossil fuels had not grown in lockstep, as before, but had (save for one markedly cold winter when fuel consumption spiked) consistently declined.

Chris Goodall and a number of other commentators have documented this decoupling extensively: U.K. government data also show a reduction in material use from about 12 tons a year per person to around 9 tons from 2000 to 2013. Japan shows a similar pattern.

Some people have contested these findings, of course. (Other people believe they are true but wish that they weren't widely known.) But there is enough evidence worldwide to show that patterns of consumption and status-seeking do change, and that intangible goods are replacing physical ones in many domains. Not only are there the obvious, comparatively trivial examples—music and film downloads, say, have replaced CDs and DVDs—but car mileage seems to have peaked, as have car purchases. Astoundingly to anyone who has seen *American*

Graffiti, half of U.S. eighteen-year-olds do not have a driver's license.

There seem to be multiple forces at work, all aligned toward a lower emphasis on material consumption. One of them may be simple satiety—it is difficult to see the benefits of not owning a car until you have owned one for a few years; it is only by traveling long-haul a few times that you may discover that your favorite place to spend your free time is a lake sixty miles from home. Now that jet travel is affordable for most people, it is perfectly acceptable (in fact rather an ornament) in wealthy British circles to take your main holiday in Britain.

Hipsterization of various categories (beer, gin, coffee, etc.) is also evidence of a complementary trend—people seeking value and status in increasingly hair-splitting distinctions between basic goods rather than spending discretionary income on greater quantities of such goods, or on non-essential purchases.

The evolutionary psychologist Geoffrey Miller has ingeniously attributed this change in behavior to the creation of online social media, which change the whole nature of status-signaling—to one in which sharing experiences may have gained signaling power at the expense of possessions.

More and more economic value is being divorced from the physical attributes of a thing and resides instead in intangibles. London's most expensive street consists of terraces of houses which most wealthy Victorians would have found laughably small. It is the fashionable address that gives them their value, and living in the center of a city is now deemed more fashionable than living in suburbia.

The great thing about intangible value, I suppose, is that its creation involves very little environmental damage. It may help disabuse people of the belief that the only way to save the planet is for us to impoverish ourselves. What it may mean is that those human qualities of status rivalry and novelty seeking which can be so destructive might be redirected, even if they cannot be eliminated.

SCIENCE MADE THIS POSSIBLE

BRUCE PARKER

Visiting Professor, Center for Maritime Systems, Stevens Institute of Technology; author, *The Power of the Sea*

There is a dichotomy in the scientific news stories we see most frequently today. The first group includes stories about exciting and/or useful scientific developments; the second includes stories about actions taken in blatant defiance of scientifically acquired knowledge and full of skepticism as to science's value. Stories from this second group include: parents refusing to vaccinate their children, the scare over genetically modified foods, ᴬᴹᴱᴺ the movement against teaching evolution in schools, skepticism about global warming, and a fear of fluoridating water.

One would think that the first group, along with other scientific writing aimed at a general audience, should work to reduce the number of news stories in the second group. Unfortunately this does not seem to be happening. In this modern era of the Internet and cable TV, the quantity of positive scientific news has certainly increased, but so has the quantity of negative scientific stories. Doubtless the positive stories are primarily read by those who already believe in science and the negative stories by those susceptible to the anti-science message. We need to find ways of bringing well-explained stories from the first group to a broader audience. And we need well-explained rebuttals to counter the stories in the second group.

Skepticism is understandable, considering the complexity

of some of the science that people are being asked to believe. Climate change/global warming is a good example. It involves physics, chemistry, biology, and geology and relies heavily on proxy data (based on isotope ratios of various elements captured in ice cores, sediment cores, corals, etc.) to provide us with records of temperature, carbon dioxide, methane, sea level, ice sheets, and other parameters needed to describe change (including ice ages) over millions of years. Such data are critical for validating computer climate models that predict future climate change. We try to explain to the non-scientist, as best we can, how the climate system works (likewise, how vaccines work, why genetically modified food is as safe as food derived by Mendel-based breeding methods, why fluoridated water is safe, how natural-selection-based evolution works, and so on). But whether or not we succeed, we must at least get across the vast amount of meticulous work and theory testing carried out by thousands of scientists to reach these conclusions.

How can we use the positive science news stories of the first group to change the minds of the audience for the second group? If the skeptics could be made to realize that their transportation, their means of communication, everything in their homes, at their jobs, and in between came originally from science, would that make a difference in their thinking? Would they be willing to look at a subject a bit more objectively? Many (most?) might still be too heavily influenced by religious beliefs or propaganda from special-interest groups to change their minds; in this Internet cable-news driven world, the negative influences are everywhere. But in writing scientific news, more emphasis should be put on the ways in which science has made modern life possible. If only we could put "Science made this possible" at the end of every scientific story, every technology story, and every story about our everyday activities. If only we could put "Science made this possible" signs on every appliance, drug, car, computer, game machine, and other such necessities of life! That might eventually make a difference.

THE BRAIN IS A STRANGE PLANET

DUSTIN YELLIN
Artist; founder, Pioneer Works

The brain is a strange planet; the planet is a strange brain. The most important science story of the past year isn't one story but an accumulation of headlines. With advanced recording and communication technologies, more people than ever before are sharing details of individual experiences and events, making culture more permeable and fluid. This verifiable record of diversity has brought people together while also amplifying their differences.

Marginalized groups are enjoying widespread recognition of their rights. Measures like federal legalization of gay marriage and partial state-level legalization of marijuana show that lifestyles once considered abnormal have gained acceptance. At the same time, political divisions remain alarmingly stark. In the past year, over 1 million fled their home countries for Europe, seeking a better life. And not all signs show we are becoming more tolerant. One of the frontrunners for the Republican presidential nomination suggested America address the threat of global terror by denying all Muslims entry into the United States, after which he performed well in the polls.

At the summit on climate change in Paris last December, delegations from India and China objected to measures that would limit their economic development by curtailing pollution. Without recognizing that today's climate is a product of damaging methods of expansion, India and China argue that emerging markets should be able to destroy the environment as did Western economies as they matured. This view is driven by fear of being usurped, outpaced, and overcome, and its myopia demonstrates that fear short-circuits the logical decision making of entire countries just as it does that of individuals.

"One person's freedom ends where another's begins," we seem to be saying. If this is what drives the formation of cultural norms today, dialogue about issues like water rights, energy use, and climate change will be determined by how those issues affect individuals.

According to F. Scott Fitzgerald, "The test of a first-rate intelligence is the ability to hold two opposed ideas in the mind at the same time and still retain the ability to function." Our culture has potential to realize a seismic shift in consciousness and rebalance environmental and social scales. We've invented the story of this world—the cities we live in, the language and symbols we use to articulate thoughts, the love we nurture to propel us forward. A shift in perspective may be all it takes to convince us that the greatest threat to humanity comes not from another people in another land but from all people, everywhere.

THE ABDICATION OF SPACETIME

DONALD D. HOFFMAN

Professor of cognitive science, UC Irvine; author, *Visual Intelligence*

Space and time have been cynosures of science at least since Einstein published his general theory of relativity in 1915, transforming them from a passive stage for the play of matter into a riveting headliner of the entire production. From the Off-Broadway venue of science, they leaped into headline news in 1919 with Eddington's confirmation during a solar eclipse that they bend, stretch, and twist, taking matter and light along for the ride. The *New York Times* headline of November 10th read: "LIGHTS ALL ASKEW IN THE HEAVENS: MEN OF SCIENCE MORE OR LESS AGOG OVER RESULTS OF ECLIPSE OBSERVATIONS."

Space and time capture the imagination precisely because they engender, and also imprison, our imagination. Imagine a holiday in Hawaii or a new design for a car, recall the wedding of a dear friend, contemplate the last moments of Custer's last stand, and in each case space and time are your helpful, even essential, partners. But then try to imagine a world of four dimensions—up/down, forward/backward, left/right, and, say, nim/zur. No one succeeds. Our partner turns jailor and straitjackets the imagination. Now try two dimensions of time, or no time at all. The straitjacket tightens.

In 1926 a brash talent debuted. Quantum theory can, in special cases, get on well with space and time, and the result of their collaboration is the Standard Model of particle physics, which successfully describes the electromagnetic, weak, and strong nuclear interactions and their associated subatomic particles. But when the density of matter is too large or the distance of interaction is too small, the collaboration breaks down and quantum theory, it now appears, can upstage its costar.

Hints of the breakdown surfaced in 1935 when Einstein, Podolsky, and Rosen observed that according to quantum theory, measurement of the quantum state of one particle can instantly change the state of another particle entangled with it, no matter how distant in space. Entanglement cannot transmit information faster than light. Nevertheless its insouciance about space and time deeply troubled Einstein.   ⌐ indifference

The breakdown splashed front and center in string theory. Nobel Laureate David Gross observed, "Everyone in string theory is convinced . . . that spacetime is doomed. But we don't know what it's replaced by." Fields medalist Edward Witten also thought that space and time may be "doomed." Nathan Seiberg, of the Institute for Advanced Study at Princeton, said, "I am almost certain that space and time are illusions. These are primitive notions that will be replaced by something more sophisticated."

The good news is that sophisticated replacements might be on the way. One new candidate is entanglement itself. Brian Swingle and Mark Van Raamsdonk found that curved spacetimes obeying Einstein's general theory of relativity can emerge from tensor networks of entangled quantum bits. In this scenario, the insouciance of entanglement is feigned. Entanglement itself is somehow the fabric that holds spacetime together.

Another new candidate is a class of geometric constructions outside space and time, including the amplituhedron discovered by Nima Arkani-Hamed and Jaroslav Trnka. Subatomic particles collide and scatter in a multitude of ways, and physicists have for decades had formulas for computing their probabilities— formulas that assume physical processes evolving locally in space and time. But as it happens, these formulas are unnecessarily complex and hide deep symmetries of nature. The amplituhedron simplifies the formulas, exposes the symmetries hidden by spacetime, and in the process abandons the assumption that space and time are fundamental.

What is fundamental, if not space and time? No one is yet sure. The prime suspect is quantum information—quantum bits and quantum gates. But quantum information viewed abstractly, not as embedded in spacetime. Spacetime and objects somehow emerge from nonspatial and nontemporal dynamics of quantum information. As John Wheeler put it, "It from bit." But this raises its own questions. Why should information, quantum or otherwise, be the bedrock of reality? And in what sense is it information?

It may be premature to write the obituary of space and time. The report of their death might be an exaggeration. But either way, dead or alive, it will be news that is important and lasting. Whether space and time prove fundamental or not, the proof itself will bring in its wake new and deep insights into the nature of reality, and perhaps also into the nature of our own imagination.

I suspect that the report of their death is not an exaggeration. This will raise new questions for researchers in perceptual psychology. Why have our perceptual systems evolved to present us a world in the format of space and time if, as Seiberg says, space and time are illusions, primitive notions that will be replaced by something more sophisticated? What selection pressures favored the ascendancy of this primitive format? What fitness advantages does it confer?

The standard assumption in perceptual psychology is that evolution favors veridical perceptions, those that accurately describe those aspects of the environment crucial to the fitness of an organism. It is not standard to assume that the very spacetime format of our perceptions is itself non-veridical, primitive, and illusory. How will this field have to change if space and time are illusions? And how will our notions of physical causality have to change? Will these changes affect how we approach the classic mind/body problem, the question of how our conscious experiences are related to our physical bodies and in particular to the activity of our brains?

Such questions make clear that the stakes are high. The grand entrance of space and time a century ago made world headlines. Their denouement will be no less riveting.

THE NEWS THAT WASN'T THERE

ANTONY GARRETT LISI

Theoretical physicist; independent researcher

On July 4, 2012, the European Organization for Nuclear Research (CERN) announced the discovery of the Higgs boson. While this was big news in fundamental physics, it was not surprising. The existence of the Higgs boson, or something like it, was necessary for the consistency of the Standard Model of particle physics, established in the 1970s and now supported by an extraordinary amount of experimental data. Finding the Higgs was central to confirming the Standard Model. However, despite the well-deserved attention accorded to the discovery of the Higgs, this was not the biggest news. The biggest recent news in fundamental physics is what has *not* been discovered: superparticles.

The theory of supersymmetry—that all existing particles are matched by "superpartners" having opposite spins—was introduced soon after the Standard Model was established and quickly became a darling of theoretical physicists. The theory helped solve a fine-tuning problem in the Standard Model, with superparticles balancing the quantum contributions to existing particle masses and making the observed masses more natural. (Although why particles have precisely the masses they do remains the largest open question in fundamental physics.) Also, for proponents of Grand Unified Theories (GUTs), the three strengths of the known Standard Model forces more perfectly converge to one value at high energies if superparticles exist. (Although a similar convergence can be achieved more simply by adding a handful of non-super bosons.) And, finally, super-symmetry (SUSY) became a cornerstone of, and necessary to, superstring theory—the dominant speculative theory of particle physics.

One of the strongest motivations for constructing the Large Hadron Collider, along with finding the Higgs, was to find superparticles. In order for SUSY to help the Standard Model's naturalness problem, superparticles should exist at energies reachable by the LHC. During the collider's first run, the antic-ipation of superparticles at CERN was palpable—it felt as if a ballroom had been set up, complete with a banner: "Welcome home SUSY!" But superparticles have not shown up at the party. The fact that the expected superparticles have not been seen puts many theorists, including string theorists, in a scientif-ically uncomfortable position.

Imagine that you had a vivid dream last night in which you saw a unicorn in your backyard. The dream was so vivid that the next day you go into your backyard and look around, expecting to find your unicorn. But it's not there. That's the position string theorists and other SUSY proponents now find themselves in. You may claim, correctly, that even though you now know there is no unicorn in the backyard, the Bayesian expectation that the unicorn is actually hiding in the closet has increased! You are "narrowing in" on finding your unicorn! This is precisely the argument SUSY proponents are present-ing, now that the LHC has failed to find superparticles near the electroweak energy scale. Yes, it is correct that the probability that the unicorn is in the closet, and that superparticles might be found during the current LHC run, has gone up. But do you know what has gone up more? The probability that the unicorn and superparticles do not exist at all. A unicorn would be a wonderful and magical animal, but maybe it, and SUSY, and superstrings really just don't exist, and it's time to think about other animals.

NO NEWS IS ASTOUNDING NEWS

LEE SMOLIN

Theoretical physicist, Perimeter Institute, Waterloo, Ontario, Canada; author, *Time Reborn*

The most important news from 2015 in fundamental physics is that probably there is no news. With one tantalizing exception (which may be a statistical anomaly), recent experiments confirm a frustratingly incomplete theory of fundamental physics which has stood since the 1970s. This is in spite of enormous effort by thousands of experimentalists hoping to discover new phenomena that would lead to greater unification and simplification in our understanding of nature.

Since 1973, our knowledge of elementary particles and fundamental forces has been expressed in what we call the Standard Model of elementary-particle physics. This reduces all phenomena, save gravity, to twelve fundamental particles interacting via three forces. The Standard Model has been confirmed in all experiments to date; this includes measurements announced in December by two teams of experimentalists operating the ATLAS and CMS detectors at the Large Hadron Collider, which is working at nearly twice the energy as in previous experiments.

In 2012, the news from the LHC was the discovery of the Higgs, the last particle predicted by the Standard Model remaining to be discovered. But the Standard Model cannot be the whole story, in part because it involves twenty-nine free parameters. We have no explanation for the values of these parameters and hence seek a deeper theory that would explain them. Moreover, many of these values seem extremely unnatural: They are very tiny numbers with large ratios among them (the hierarchy problem), and they seem to be tuned to special values needed for a universe with many stable nuclei allowing complex life to exist (the fine-tuning problem). In addition, there's no reason for the choices of the fundamental particles or the symmetries governing the forces between them. Another reason for expecting new particles beyond the Standard Model is that we have excellent evidence from astronomy for dark matter, which gravitates but doesn't give off light. All these pieces of evidence point to new phenomena that could have been discovered at the LHC.

Several beautiful hypotheses have been offered on which to base a deeper unification. I'll just give the names here: supersymmetry, technicolor, large extra dimensions, compositeness. These each imply that the LHC should have discovered new particles. Some also point to more exotic phenomena, such as quantum black holes. To date, the experimental evidence sets impressive limits against these possibilities.

To be sure, there is one weak but exciting indication from new results that might be interpreted as signaling a new particle beyond the Standard Model. This is a small excess of collisions which produce pairs of photons that, remarkably, are seen by both of the experiments operating at the LHC. But the statistical significance is not high, given that we are bound to get some signal by random chance in one of the many channels looked at. So this could be a random fluctuation that will go away when more data are taken.

Even if this hint grows into the discovery of a new particle, which would be extremely exciting news, it is too soon to say whether it will lead to a deeper unification or just add complication to the already complicated Standard Model. Luckily, more data can be expected soon.

It's the same with quantum gravity, the unification of quantum theory with Einstein's theory of gravity. Many proposals for quantum gravity suggest that at certain very high energy scales we must see new physics. This would indicate that at correspondingly tiny scales space becomes discrete, or new features of quantum geometry kick in. One consequence would be that the speed of light is no longer universal—as it is in relativity theory—but would gain a dependence on energy and polarization visible at certain scales.

In the last decade, this prediction has been tested by sensitive measurements of gamma rays that have traveled for billions of years from extremely energetic events called gamma-ray bursts. If the speed of light depends even slightly on energy, we would see higher-energy photons arriving systematically earlier or later than lower-energy photons; the enormous travel time would amplify the effect. This has been looked for by the Fermi Gamma-ray Space Telescope and other detectors of gamma rays and cosmic rays. No deviations from relativity theory are seen. Thus, our best hope of discovering quantum-gravity physics has been frustrated.

A similar story seems to characterize cosmology. Something remarkable happened in the very early universe to produce a world vast in scale but at the same time extremely smooth and homogeneous. One explanation for this is inflation, a sudden enormous expansion at very early times, but there are competitors. Each of these theories requires delicate fine-tuning of parameters and initial conditions. Once this tuning is done, each predicts a distribution of noisy fluctuations across the smooth universe. The fluctuations show up as a seemingly random distribution of very slightly denser and less dense regions which, over hundreds of millions of years of expansion, amplify and give rise to the galaxies. These fluctuations make bumps that are visible in the cosmic microwave background (CMB) radiation. So far, their distribution is as random, featureless, and boring as possible, and the simplest theories—whether inflation or its alternatives—suffice to explain them.

In each of these domains, we have sought clues from experiments into how nature goes beyond, and solves the puzzles latent in, our incomplete theories of the universe, but we have so far come up with nearly nothing. It's beginning to seem as if nature is just unnaturally fine-tuned. In my opinion, we should now be seeking explanations for why this might be. Perhaps the laws of nature are not static but have evolved, through some dynamical mechanism, to have the unlikely forms they are observed to have.

ONE HUNDRED YEARS OF FAILURE

SETH LLOYD

Professor of quantum-mechanical engineering, MIT;
author, *Programming the Universe*

The year 2015 marked the hundredth anniversary of Einstein's announcement of the general theory of relativity. General relativity describes the force of gravity in terms of the curvature of space and time: The presence of matter warps the underlying fabric of the universe, causing light to curve and clocks to slow down in the presence of matter. General relativity supplies us with a physical theory allowing us to describe the cosmos as a whole; it predicts the existence of such exotic objects as black holes; it even supports closed timelike curves that in principle allow travel backward in time. General relativity is a tremendous scientific success story, and its centennial was accompanied by many articles, television shows, scientific conferences, and more to celebrate Einstein's achievement.

Unmentioned in this celebration was the darker story. As soon as Einstein announced his elegant theory, other physicists began trying to reconcile general relativity with quantum mechanics. Quantum mechanics is the physical theory governing matter at its smallest and most fundamental scales. The last century has seen tremendous advances in its application to the study of elementary particles, solid-state physics, the physics of light, and the fundamental physics of information-processing. Pretty much as soon as the print on Einstein's papers had dried, physicists began trying to make a quantum theory of gravity. They failed.

The first theories of quantum gravity failed because scientists did not understand quantum mechanics very well. It was not until a decade after Einstein's results that Erwin Schrödinger and Werner Heisenberg offered a precise mathematical formulation of quantum mechanics. By the beginning of the 1930s, Paul Dirac had formulated a version of quantum mechanics that incorporated Einstein's earlier—and by definition less general—special theory of relativity. Throughout the next half century, in the hands of physicists such as Richard Feynman and Murray Gell-Mann, this special-relativistic version of quantum mechanics, called quantum field theory, provided dramatic advances in our understanding of fundamental physics, culminating by the mid 1970s in the so-called Standard Model of elementary particles. The Standard Model unifies all the known forces of nature apart from gravity: It has been confirmed by experiment again and again.

What about a quantum theory of gravity, then? After Dirac, when physicists tried to extend the successful techniques of quantum field theory to general relativity, they failed. This time they failed because of a knotty technical problem. One of the peculiarities of quantum field theory is that when you try calculating the value of some observable quantity, such as the mass of the electron, the naïve answer you obtain is infinity.

Looking more closely, you realize that the interactions between the electron and other particles (such as photons, particles of light) have to be taken into account: These interactions "renormalize" the mass of the electron, making it finite. Renormalization works beautifully in the case of quantum field theory, allowing the prediction of quantities such as the mass of the electron to more than six digits of accuracy. But renormalization fails utterly in the case of quantum gravity:

Quantum gravity is not renormalizable. Infinity remains infinity. Failure.

More recent decades of failure to quantize general relativity have yielded tantalizing clues. Perhaps the best-known result combining quantum mechanics and gravity is Stephen Hawking's famous proof that black holes are not absolutely black but in fact emit radiation. Hawking radiation is not a theory of quantum gravity, however, but of quantum matter moving about on a classical spacetime that obeys Einstein's original non-quantum equations. Loop quantum gravity solves some of the problems of quantum gravity, but exacerbates others: It has a hard time including matter in the theory, for example. Speaking as someone made of matter, I object to theories that do not include it.

One of the primary appeals of string theory is that it naturally contains a particle that could be identified with the graviton, the quantum of gravity. Sadly, even the most enthusiastic followers of string theory admit that it is not yet a fully self-consistent theory but a series of compelling mathematical observations called—with an apparent lack of irony—"miracles."

Longtime practitioners of quantum gravity have advised me that if one wishes to publish in the field, any advance that claims to improve on one aspect of quantum gravity must be offset by making other problems worse, so that the net effect is negative. If economics is the dismal science, then quantum gravity is the dismal physics.

The last few years have seen a few glimmers of hope, however. Quantum information is the branch of physics and mathematics describing how systems represent and process information in a quantum-mechanical fashion. Unlike string theory, quantum information is in fact a theory: It proceeds by orderly conjecture and mathematical proof, with close contact to experiment.

Quantum information can be thought of as the universal theory of discrete quantum systems—systems that can be represented by bits, or qubits (quantum bits).

Recently, researchers in quantum gravity and quantum information have joined forces to show that quantum-information theory can provide deep insights into problems such as black-hole evaporation, the holographic principle, and the AdS-CFT [anti-de Sitter/conformal field theory] correspondence. (If these subjects sound esoteric, that's because they are.) Encouragingly, the advances in quantum gravity supplied by quantum-information theory do not yet seem to be counterbalanced by backsliding elsewhere.

We have no idea whether this attempted unification of qubits and gravitons will succeed or fail. Empirical observation of the last century of failure to quantize gravity suggests the latter. With any luck, however, the next hundred years of quantizing gravity will not be so dismal.

HOPE BEYOND THE HIGGS BOSON

SARAH DEMERS

Horace Taft Associate Professor of Physics, Yale University

Imagine that a friend you trust tells you a rumor. It's an unlikely story and they aren't completely sure of themselves. But a few minutes later, another friend cautiously tells you the same thing. The combination of two similar stories from two reputable witnesses makes you want to explore further. This is what happened on December 15, 2015, when the ATLAS and CMS experiments at CERN announced an initial analysis of their highest-energy run, with the same hint of something interesting in the data. It's too early to claim a new particle, but the situation in particle physics makes this the biggest news in our recent history.

The Large Hadron Collider's energy-breaking run was launched with a vengeance in 2010, following an incident that damaged the machine in 2008 and a cautious year in 2009. The 2011–2012 data set delivered the discovery of the Higgs boson. As data streamed in, particle physicists around the world clustered in conversation around espresso machines on this sobering scenario: What if we find the Higgs boson and nothing else? In other words, what if we neatly categorize the particles predicted by the Standard Model, incorporating the Higgs mechanism to provide mass, but make no progress toward understanding the nature of dark matter, dark energy, quantum gravity, or find any clues to explain the 96 percent of the matter/energy content of the universe which isn't incorporated in the theory?

If physics is a valid framework for understanding the universe, there's something else out there for us to discover. The prediction of the Higgs, the ultimately successful decades-long campaign to discover it, and the ongoing partnership between experiment and theory to characterize it, gives us confidence in the methods we're using. But the missing pieces could be beyond our imaginations, the current state of our technology, or both.

The terrifying possibility floating through these "Higgs and nothing else" conversations is that we might reach the end of exploration at the energy frontier. Without better clues to our undiscovered physics, we might not have sufficient motivation to build a higher-energy machine. Even if we convince ourselves, could we convince the world and marshal the necessary resources to break the energy frontier again and continue probing nature under the extreme conditions that teach us about nature's building blocks?

In 2015, the LHC broke another energy barrier for hadron colliders with a jump from the 8 TeV center-of-mass collisions that produced the Higgs to 13 TeV center-of-mass. The ATLAS and CMS collaborations worked 24/7 to analyze the data in time for a presentation at an end-of-year event on December 15. Both experiments cautiously reported the hint of a new particle with the same signature in the same place. Two photons caught in high-energy collisions can be arranged together to form a mass, as if they originated from the same single particle. Using this technique, both experiments saw a slight clustering of masses—more than what was expected from the Standard Model alone—near 750 GeV/c^2. The experimenters are cautious for a good reason. These hints, in the same place for ATLAS and CMS, could disappear with the gathering of data in 2016. If the situation were less dire, this would not be big news. But for me it represents the promise of our current energy-frontier physics program.

Over the next few years, we'll gather ten times the data at 13 TeV than we currently have, and we have theoretical reasons to expect something right around the corner. If the 750 GeV hint disappears with increased statistics, we'll keep searching for the next hint that could break open our understanding of nature. We move forward with the next step potentially within our reach, determined to find it, if it's there.

AN UNEXPECTED, HAUNTING SIGNAL

GERALD HOLTON

Mallinckrodt Research Professor of Physics, professor of the history of science, emeritus, Harvard University; author, *Victory and Vexation in Science: Einstein, Bohr, Heisenberg, and Others*

The big news for many in the physics community worldwide last December was that at the collider at CERN an unexpected, haunting signal of a possible new particle was found, one that would not fit into any part of the current high-energy theory. An entirely new region in experiment and theory might open up, making the finding of the Higgs an old story.

But just after this finding, the collider was shut down, as usual, until spring. There ensued a tantalizing wait, in a way analogous to the wartime event when the first nuclear reactor, in Chicago, under Enrico Fermi, got to the brink of criticality late one morning. But instead of continuing, Fermi asked everyone there to wait until he would return from his midday siesta.

NEWS ABOUT HOW THE PHYSICAL WORLD OPERATES

LEONARD SUSSKIND
Felix Bloch Professor of Theoretical Physics, Stanford University;
co-author (with Art Friedman), *Quantum Mechanics: The Theoretical Minimum*

I'll try to report the news from the physics front that I think may prove to be important. When I say "important," I mean to someone interested in how the physical world operates.

First of all, from the experimental front there is news from the CERN Large Hadron Collider—evidence of a new particle. What "new particle" means at this stage is a small bump in a data distribution. It could be real or it could be a statistical fluke, but if real it does represent something new. Unlike the Higgs particle, it is not part of the Standard Model of particle physics. In fact, to my knowledge the new particle does not fit neatly into any theoretical framework, such as supersymmetry or technicolor, and it's not a black hole or a graviton. So far, it just seems to be an extra particle.

If it's real and not just a fluke, then there will probably be more particles uncovered, and not only new particles but new forces—perhaps a whole new structure on top of the Standard Model. At the moment, no one has a compelling idea of what it means. It might be connected to the puzzle of dark matter, the missing matter in the universe that seeded the galaxies. The new particle is not itself dark matter—it's too short-lived—but other, related particles could be.

From the more theoretical side, what I find most interesting is new ideas that relate gravity, the structure of space, and quantum mechanics. For example, there is gathering evidence (all theoretical) that quantum entanglement is the glue holding space together. Without quantum entanglement, space would fall apart into an amorphous, unstructured, unrecognizable thing.

Another idea (Full disclosure: It's my idea) is that the emergence of space behind the horizons of black holes is due to the growth of quantum complexity. This is too technical to explain here, except to say that it's a surprising new connection between physics and quantum-information science. It's not a completely far-fetched idea that these connections may not only teach us new things about fundamental physics problems but also be tools for understanding the more practical issues for building and using quantum computers. Stranger things have happened.

UNPUBLICIZED IMPLICATIONS OF HAWKING BLACK-HOLE EVAPORATION

FRANK TIPLER

Mathematical physicist, cosmologist, Tulane University;
author, *The Physics of Immortality*

In 1974, Stephen Hawking proved that black holes were not black. Rather, quantum mechanics required that black holes would slowly lose their mass via a really neat mechanism: The gravity of a black hole would create a pair of particles outside the black hole, one particle with negative mass and the other with positive mass; the former would fall inside the black hole and the latter would move away from the black hole. The net effect would be to decrease the mass of the black hole, and its mass would eventually go to zero.

Hawking realized that a zero-mass black hole was a big no-no, because such an entity could only be a naked singularity that destroyed the information inside the black hole. One of the fundamental principles of quantum mechanics—the same theory that tells us black holes evaporate—is "unitarity," which means that information is conserved. But if a black hole destroys information in the final stages of its evaporation, then information cannot be conserved. Unitarity would be violated if a black hole were to evaporate completely.

Hawking then made a mistake: he argued that we have to accept a violation of unitarity in black-hole evaporation. But unitarity is a fundamental principle of quantum physics. Unitarity has many implications, one of which is that if unitarity is violated, so is the conservation of energy. And a little violation of unitarity is like being "just a little pregnant"; it has a tendency to get larger—very much larger. Leonard Susskind of Stanford University pointed out that a tiny violation of unitarity would give rise to a disastrous positive feedback of violation of energy conservation: If one were to turn on a microwave oven, so much energy would be created out of nothing—conservation of energy does not hold, remember—that the Earth would be blown apart!

Obviously this cannot happen. Information, and hence energy, must be conserved. But many black holes have been detected, and they must evaporate. How is this dilemma to be resolved?

There is an obvious resolution: namely, that all observed black holes are the mass of the Sun or larger and such black holes will last billions of trillions of years before they approach a naked singularity. What if the universe came to an end in a Big Crunch singularity before any black holes had time to evaporate completely?

This resolution of the black-hole evaporation dilemma has a host of fascinating implications. First, it means that the dark energy, whatever it is, will eventually turn off. In an ever-accelerating universe, there will be no Big Crunch singularity, and all black holes will eventually evaporate.

Second, the great Israeli-American physicist Jacob Bekenstein—whose work suggested to Hawking that he should investigate the possibility of black-hole evaporation—has proved mathematically that if event horizons exist, then the entropy of the universe must approach zero as a Big Crunch singularity is approached. But the second law of thermodynamics says that entropy can never decrease, much less approach zero at the end of time. Thus, if the second law holds forever—which it does—then event horizons cannot exist. The absence of event horizons can be shown mathematically to imply that the universe must be spatially finite.

The absence of event horizons also incidentally, almost in passing, resolves the problem of how the information inside a black hole escapes: If there are no event horizons, there is no barrier to getting out. We should keep in mind that the assumption that a black hole is bounded by an event horizon is just that—an assumption, not an observed fact. If the universe ends in a Big Crunch, the information inside a black hole would not get out until near the Big Crunch. No observation today can show that the information is forever bound to being inside the black hole. A claim that event horizons exist is like someone's claim to be immortal: You would have to wait until the end of time to confirm the claim.

So by merely accepting the obvious resolution of Hawking's dilemma and applying the standard laws of physics, we infer that the universe is spatially finite, that the dark energy will eventually turn off, that the universe will end in a Big Crunch, and that event horizons do not exist. Various physicists have pointed to each of these facts over the past decade, but the implications seem to have escaped the science journalists. Eventually the information will leak out, hopefully before the end of time.

THE ENERGY OF NOTHING

ANDREI LINDE

Author of eternal chaotic inflation; theoretical physicist, Stanford University

Back in 1998, two groups of astrophysicists studying supernovae made one of the most important experimental discoveries of the 20th century: They found that empty space, the vacuum, is not entirely empty. Each cubic centimeter of empty space contains about 10^{-29} grams of invisible matter—or, equivalently, vacuum energy. This is almost nothing: 29 orders of magnitude smaller than the mass of matter in a cubic centimeter of water, 5 orders of magnitude less than the mass of a proton. If the Earth were made of such matter, it would weigh less than a gram.

If the vacuum energy is so small, how do we even know it's there? Just try putting 10^{-29} grams on the most sensitive of scales; it will show nothing at all. At first, many people were skeptical about its existence, but the combined efforts of cosmologists who study cosmic microwave background radiation and large-scale structure of the universe not only confirmed this discovery but allowed measurement of the energy density of the vacuum within a few-percent accuracy. Doubts and disbelief were replaced by acceptance, and by the Nobel Prizes received in 2011 by Saul Perlmutter, Brian Schmidt, and Adam Riess.

The news has shaken physicists all over the world. But is it much ado about nothing? If something is so hard to find, maybe it's irrelevant and not at all newsworthy.

The vacuum energy is extremely small indeed, but in fact it

is comparable to the average energy density of normal matter in the universe. Before this discovery, astronomers believed that the density of matter constituted only 30 percent of the density corresponding to a flat universe. That meant that the universe is open, contrary to the prediction of the inflationary theory of the universe's origin. The vacuum energy added the required 70 percent to the sum total, thus confirming one of the most important predictions of inflationary cosmology.

The tiny vacuum energy is large enough to make our universe slowly accelerate. It will take about 10 billion more years for the universe to double in size, and if this expansion continues, in about 150 billion years all distant galaxies will forever disappear from our view. This constitutes quite a change from our previous expectation that in the future we would see more and more. . . .

The possibility that the vacuum may have energy was discussed almost a century ago by Einstein, but then he discarded the idea. Particle physicists reintroduced it, but their best estimates of its density were far too large to be true. For a long time they tried to find a theory explaining why vacuum energy must be zero, but all such attempts failed. Explaining why it is not zero but incredibly, excruciatingly small is a much greater challenge.

There is an additional problem: As noted, vacuum energy is at present comparable with the average energy density of matter in the universe. In the past, the universe was small, and its vacuum energy was negligibly small compared to the energy density of normal matter. In the future, the universe will grow, and the density of normal matter will become exponentially small. Why do we live exactly at the time when the energy of empty space is comparable to the energy of normal matter?

Thirty years ago, well before the discovery of the energy of nothing, Steven Weinberg and several other scientists argued that observing a small value of the vacuum energy would not be too surprising: A universe with a large negative vacuum energy would collapse before life had a chance to emerge, whereas a large positive vacuum energy would not allow galaxies to form. Thus we could live only in a universe with a sufficiently small absolute value of vacuum energy. But this anthropic argument, by itself, was not sufficient. We used to think that all parameters of the theory of fundamental interactions, such as vacuum energy, are just numbers that are given to us and cannot change—which is why the vacuum energy was also called the cosmological constant. But if the vacuum energy is a true constant that cannot change, anthropic considerations cannot help.

The only known way to solve this problem is in the context of the theory of an inflationary multiverse and string-theory landscape, which claims that the universe consists of many parts with different properties and different values of vacuum energy. We can live only in those parts where the vacuum energy is small enough, which explains why the vacuum energy is so small in the part of the world in which we live.

Some people are critical of this way of thinking, but in the eighteen years since the discovery of the vacuum energy, nobody has come up with a convincing alternative solution to the problem. Many others are excited, including Steven Weinberg, who exclaimed, "Now we may be at a new turning point, a radical change in what we accept as a legitimate foundation for a physical theory!"*

This explanation of a small vacuum energy has an unexpected twist to it: According to this scenario, all vacua of our

---

* "Living in the Multiverse," arXiv.org/abs/hep-th/0511037 (2005).

type are not stable but metastable. This means that in a distant future our vacuum will decay, destroying life as we know it in our part of the universe while re-creating it over and over again in other parts.

It is too early to say whether these conclusions are here to stay or will be significantly modified in the future. In any case, it is amazing that the news of a seemingly inconsequential discovery of an incredibly small energy of empty space may have enormous consequences for cosmology, string theory, scientific methodology, and even for our view of the ultimate fate of the universe.

THE BIG BANG CANNOT BE WHAT WE THOUGHT IT WAS

PAUL J. STEINHARDT

Theoretical physicist; director, Princeton Center for Theoretical Science, Princeton University; co-author (with Neil Turok), *Endless Universe*

Two years ago, the scientific community and the press trumpeted the claim by a team of scientists that they had found definitive proof that the universe began with a Big Bang followed by a period of accelerated expansion known as inflation. Their proof was that the light produced in the infant universe and collected by their detectors exhibited a distinctive pattern of polarization that could be explained only if the large-scale structure of the universe was set when the temperature and density of the universe were extraordinarily high, just as posited in the Big Bang inflationary picture.

Over the ensuing year, though, it became clear that the claim was a blunder. In searching for a cosmic signal from the distant universe, the team had not taken proper account of the polarization of light that occurred nearby when it passed through the dust in our Milky Way on the way to their detectors. The new claim from the team, published in recent months, is that there is no sign of the cosmic polarization they had been seeking despite an extensive search with extraordinarily sensitive detectors.

The retraction received considerable attention, but the full import of the news has not been appreciated: *We now know that the Big Bang cannot be what we thought it was.*

The prevailing view has been that the Big Bang was a violent, high-energy event, during which space, time, matter, and energy were suddenly created from nothing in a distorted, nonuniform distribution. To account for the undistorted nearly uniform universe we actually observe, many cosmologists hypothesize a period of rapid stretching (inflation) just after the Bang, when the concentration of energy and matter was still very high. If there were inflationary stretching only, the universe would become perfectly smooth, but there is always quantum physics in addition to stretching, and quantum physics resists perfect smoothness.

At the high concentrations of energy required for inflation, random quantum fluctuations keep generating bumps and wiggles in the shape of space and the distribution of matter and energy—irregularities that should remain when inflation ends. The quantum-generated irregularities should appear today as hot spots and cold spots in the pattern of light emanating from the early universe—the so-called cosmic background radiation. The hot and cold spots have indeed been observed and mapped in numerous experiments since the COBE satellite detected the first spatial variations in the cosmic-background-radiation temperature in 1992.

The problem is that when the concentration of energy is high, the quantum-generated distortions in space should modify the way light scatters from matter in the early universe and imprint a spiraling pattern of polarization across the cosmos. It was the detection of this spiraling pattern (referred to as B-mode) that was claimed as proof of the Big Bang inflationary picture and then retracted. The failure to detect the B-mode pattern means that there is something very wrong with the picture of a violent Big Bang followed by a period of high-energy-driven inflation. Whatever processes set the large-scale structure of the universe had to be a gentler, lower-energy process than has been supposed.

Simply lowering the energy concentration at which inflation starts, as some theorists have suggested, only leads to more trouble. This leaves more time after the Big Bang for the nonuniform distribution of matter and energy to drive the universe away from inflation. Starting inflation after the Big Bang and having enough inflation to smooth the universe becomes exponentially less likely as the energy concentration is lowered. The universe is more likely to emerge as too rough, too curved, too inhomogeneous compared to what we observe.

Something more radical is called for. Perhaps an improved understanding of quantum gravity will enable us to understand how the Big Bang and inflation can be discarded in favor of a gentler beginning. Or perhaps the Big Bang was actually a gentle bounce from a previous period of contraction to the current period of expansion. During a period of slow contraction, it is possible to smooth the distribution of space, matter, and energy and to create hot spots and cold spots without creating any B-modes at all.

As the news sinks in, scientists will need to rethink, depending on whether forthcoming more sensitive efforts to detect a B-mode pattern find anything at all. Whatever is found, our view of the Big Bang will be changed, and that is newsworthy.

ANOMALIES

STEPHON H. ALEXANDER

Theoretical physicist, Dartmouth College; author, *The Jazz of Physics*

These days, physics finds itself in a situation similar to that faced by physicists at the turn of the 20th century, just before the dawn of quantum mechanics and general relativity. In 1894, the codiscoverer of the constancy of the speed of light and the first American Nobel laureate in physics, Albert Michelson, stated, "It seems probable that most of the grand underlying principles [of physical science] have been firmly established." Many, perhaps most, physicists at that time believed that the handful of experimental anomalies that seemed to violate those principles were minor details that would eventually be explained by the paradigm of classical physics. Within a generation, quantum mechanics and Einstein's theory of relativity were invented and classical physics overthrown, in order to explain those minor experimental details.

In 2012, I spent a year on sabbatical at Princeton, at the invitation of David Spergel, one of the lead scientists of the WMAP [Wilkinson Microwave Anisotropy Probe] space satellite. WMAP had been designed to make the most precise measurements of the ripples in the cosmic microwave background radiation, the afterglow of the Big Bang—ripples that, according to our Standard Model of cosmology, would develop into the vast structures in our universe, to which there is remarkable agreement. Cosmic inflation, our best theory of the early universe, is consistent with the most precise physics known to us, general relativity and quantum field theory.

Despite the success of inflation in predicting the features observed by WMAP, nagging anomalies persist—just as was true at the end of the 19th century. My Princeton colleagues and I spent that year wrestling with those anomalies to no avail. David comforted me with the idea that the WMAP anomalies were probably due to an unaccounted-for experimental factor rather than some strange phenomenon in the sky. "If the anomalies persist with the *Planck* satellite," he warned, "we will have to take them more seriously." Well, in 2014 *Planck* made an even more precise measurement of the CMB anisotropies and presented us with arguably the most nagging in the suite of anomalies: the hemispherical anomaly.

What's that?

The undulations in the CMB reflect a prediction from cosmic inflation that aside from those tiny waves, on the largest distance scales, the universe looks the same in every direction and at every vantage point in space. This prediction is consistent with one of the pillars of modern cosmology, the cosmological Copernican principle. During the epoch when the CMB anisotropies were formed, they too were supposed to (on average) look the same in every direction. The theory of cosmic inflation, in which rapid expansion of spacetime smooths out any large-scale directional preference while democratically sprinkling the spacetime fabric with the same amount of ripples in every direction, predicts this feature. This means that if one divides the sky into two arbitrary hemispheres, we should see the same statistical features of the anisotropies in both hemispheres.

However, both WMAP and *Planck* see a difference in the

amount of anisotropies in the hemispheres. With some tweaking, it is possible to modify inflation to account for the anomaly, but this seems at odds with what inflation was invented for—to make the early universe smooth and allow for the tiny anisotropies that later become galaxies. One might think this would provide opportunity for alternate theories of the early universe, such as bouncing/cyclic cosmologies, to rise to the occasion and explain the anomaly, but so far there is no compelling explanation.

Recently, at the Large Hadron Collider in Geneva, the ATLAS and CMS experiments both reported an anomaly when protons collided on energy scales close to 1 trillion electron volts. The experiments saw the predicted production of elementary particles that the standard quantum field theory of elementary-particle interactions predicts, with the exception of an excess production of light. If this observation persists with statistical significance, we will need physics beyond our Standard Model to explain.

To me and my theoretical colleagues, both anomalies are a good thing. Will they lead to simple yet less-than-pretty fixes of our current Standard Models? Might one of our supertheories—supersymmetry, strings, loop quantum gravity, GUTs—come to the rescue? Could it be that the anomalies are connected in some yet unseen way? Maybe they will point us in an entirely unthought-of direction. Whatever the case, this is an exciting time to be a theorist.

LOOKING WHERE THE LIGHT ISN'T

BRIAN G. KEATING

Professor, Department of Physics and the Center for Astrophysics and Space Sciences, UC San Diego

For decades, the search strategy of confining the hunt for your lost keys to beneath the streetlight has been employed both by drunks and neutrino hunters, with no keys in sight and few key insights.

In 1916, Einstein published his final general relativity paper. One hundred years later, using Einstein's predictions, we are on the brink of "weighing" the last elementary particle whose mass is unknown. Isn't this old news? Don't we know all the fundamental particle masses already, after measuring the Higgs boson's mass? Well, yes and no.

Looking at the Standard Model, we see sixteen subatomic particles: quarks, leptons (such as the electron), and bosons (such as the photon), plus the Higgs boson, charted in a table reminiscent of Mendeleev's periodic table of the elements—except that there is no periodicity, no apparent ordering at work.

Three of the six leptons ("small" in Greek; particles that don't participate in the strong nuclear force) are the three "generations" of neutrinos: electron, muon, and tau neutrinos. As integral as they are to the foundations of matter, we're in the dark about their masses. A particle's mass is arguably its most distinctive property, so this lacuna is rightly seen as an embarrassment for physics. That is about to change.

Neutrinos are generated in nuclear reactions such as fusion

and radioactive decay. The ultimate reactor, of course, was the biggest cauldron of them all: the Big Bang. Like light, neutrinos are stable. Their lifetimes are infinite, because, like light, there is nothing for them to decay into. They change their flavor (generation type) as they sail through the cosmos, a phenomenon called oscillation. The 2015 Nobel Prize in physics went to Takaaki Kajita and Arthur B. McDonald "for the discovery of neutrino oscillations, which shows that neutrinos have mass." Their work devastatingly refutes claims presented in John Updike's poem *Cosmic Gall* (Sorry, John): While they remain small, neutrinos *do* have mass after all. Thanks to Kajita and McDonald, not only do we know neutrinos have mass but their work gives us a lower limit on the masses. At least one of the three must have a mass bigger than about 1/20 of an electron volt. This is svelte; the next heaviest elementary particle is the electron, whose mass is 10 million times larger. Most important, these lower limits on neutrino masses give experimentalists thresholds to target. All that's left is to build a scale sensitive enough to weigh them.

Since it's impossible to collect enough neutrinos to weigh in a terrestrial laboratory, cosmologists will use galaxy clusters as their scales. Sprinkled amid the luminous matter in the clusters are innumerable neutrinos. Their masses can be measured using gravitational lensing, a direct consequence of Einstein's general theory. All matter, dark and luminous, gravitationally deflects light. The gravitational-lensing effect rearranges photon trajectories, as Eddington showed during the 1919 total solar eclipse. Star positions were displaced from where they would have been seen in the absence of the Sun's warping of spacetime. The light that should have been there was lensed; the amount of displacement told us the mass of the lens.

What kind of light should we use to weigh poltergeist particles like neutrinos? There certainly aren't enough neutrinos in our solar system to bend the Sun's light. The most promising light source of all is also the oldest and most abundant light in the universe, the "3 Kelvin" cosmic microwave background. These cosmic photons arose from the same ancient cauldron that produced the neutrinos plying the universe today. The CMB is "cosmic wallpaper," a background against which the mass of all matter in the foreground galaxy clusters, including neutrinos, can be measured.

In 2015, the *Planck* satellite showed powerful evidence for gravitational lensing of the CMB, using a technique eventually guaranteed to detect neutrino masses. This technique, based on the CMB's polarization properties, will dramatically improve in 2016, thanks to a suite of experiments deploying tens of thousands of detectors cooled below 0.3 Kelvin at the South Pole and in the Chilean Atacama desert.

Neutrinos are also the paradigm of dark matter: they're massive, dark (they interact with light only via gravitational lensing) and neutral—all required properties of dark matter. While we know that neutrinos aren't the dominant form of the cosmos's missing mass, they are the only *known* form of dark matter. After we measure their masses, we'll use neutrinos to thin the herd of potential dark-matter candidates. Just as there are many different types of ordinary matter, ranging from quarks to atoms, we might expect there to be several kinds of dark matter. Perhaps there is a "dark" periodic table.

The hunt is on to directly detect dark matter, and several upgrades to liquid noble-gas experiments are coming online in 2016. Perhaps there will be detections. But so far the direct-detection experiments have produced only upper limits on the mass of the dark matter other than neutrinos. In the end, neutrinos just might be the only form of dark matter we ever get to "see."

The next century of general relativity promises to be as exciting as the first. "Spacetime tells matter how to move; matter tells spacetime how to curve," said John Archibald Wheeler. We've seen what the curvature is. Now we just need to find out what's the matter. And where better to look for lost matter than where the dark is.

SIMPLICITY

NEIL TUROK

Director, Perimeter Institute, Waterloo, Ontario; Niels Bohr Chair in Theoretical Physics; author, *The Universe Within*

We live at a remarkable moment in history. Our scientific instruments have allowed us to see the far reaches of the cosmos and study the tiniest particles. In both cases, they have revealed a surprising simplicity, at odds with the most popular theoretical paradigms. I believe this simplicity to be a clue to a new scientific principle, whose discovery will represent the next revolution in physics and our understanding of the universe.

It is not without irony that at the very moment the observational situation is clearing so beautifully, the theoretical scene has become overwhelmingly confused. Not only are the most popular models very complicated and contrived, they are also being steadily ruled out by new data. Some physicists appeal to a "multiverse" in which all possible laws of physics are realized somewhere, since then (they hope) there would at least be one region like ours. It seems more likely to me that the wonderful new data are pointing us in the opposite direction. The cosmos isn't wild and unpredictable, it is incredibly regular. In its fundamental aspects, it may be as simple as an atom and eventually just as possible to understand.

Our most powerful microscope, the Large Hadron Collider, has found the Higgs boson. This particle is the basic quantum of the Higgs field, a medium that pervades space and endows particles with mass and properties like electric charge. As fundamental to our understanding of particle physics as the Higgs field is, it is equally important to our understanding of cosmology. It makes a big contribution to the energy in empty space, the so-called dark energy, whose density astronomical observations reveal to be a weirdly tiny, yet positive, number. Furthermore, according to the LHC measurements and the Standard Model of particle physics, the Higgs field is delicately poised on the threshold of instability in today's universe.

The discovery of the Higgs boson was a triumph for the theory of quantum fields, the amalgamation of quantum mechanics and relativity that dominated 20th-century physics. But quantum field theory has great trouble explaining the mass of the Higgs boson and the energy in empty space. In both cases, the problem is essentially the same. The quantized vibrations of the known fields and particles become wild on small scales, contributing large corrections to the Higgs boson mass and the dark-energy density and generally giving them values much greater than those we observe.

To overcome these problems, many theorists have postulated new particles, whose effects would almost precisely cancel those of all the known particles, "protecting" the mass of the Higgs boson and the value of the dark-energy density from quantum effects. But the LHC has looked for these extra partner particles and, so far, failed to find them. It seems that nature has found a simpler way to tame quantum phenomena on short distances, in a manner we have yet to fathom.

Meanwhile, our most powerful telescope, the *Planck* satellite, has scanned the universe on the largest visible scales. What it has revealed is equally surprising. The whole shebang can be quantified with just six numbers: the age and temperature of the cosmos today; the density of the dark energy and the dark matter (both mysterious, but simple to characterize); and the strength, and slight dependence on scale, of the tiny initial variations in the density of matter from place to place as it emerged from the Big Bang. None of the complications, like gravitational waves or the more involved density patterns expected in many models, appear to be there. Again, nature has found a simpler way to work than we can currently understand.

The largest scale in physics—the Hubble length—is defined by the dark energy. By accelerating the expansion of the cosmos, the dark energy carries distant matter away from us and sets a limit to what we will ultimately see. The smallest scale in physics is the Planck length, the minuscule wavelength of photons so energetic that two of them will form a black hole. While exploring physics down to the Planck length is beyond the capabilities of any conceivable collider, the universe itself probed this scale in its earliest moments. So the simple structure of the cosmos is likely to be an indication that the laws of physics become simple at this extreme.

All the complexity in the world—including stars, planets, and life—apparently resides in the "messy middle." It is a striking fact that the geometric mean of the Hubble and Planck lengths is the size of a living cell—the scale on which we live, where Nature is at her most complex.

What is exciting about this picture is that it requires a new kind of theory, one that is simple at both the smallest and largest scales, and at very early and very late cosmological times, so that it can explain these properties of our world. In fact, there are more detailed hints, from both theory and data, that at these extremes the laws of physics should become *independent* of scale. Such a theory won't be concerned with kilograms, meters, or seconds, only with information and its relations. It will be a unified theory not only of all the forces and particles but also of the universe as a whole.

THE LHC IS WORKING AT FULL ENERGY

GORDON KANE

Theoretical physicist and cosmologist; Victor Weisskopf Distinguished University Professor, University of Michigan; author, *Supersymmetry and Beyond*

The most interesting recent physics news is that the Large Hadron Collider at CERN, in Geneva, is finally working at its highest-ever design energy and intensity. Why that is so important is because it may at last allow the discovery of new particles—superpartners—that would enable formulating and testing a final theory underlying the physical universe.

As Max Planck immediately recognized when he discovered quantum theory over a century ago, the equations of the final theory should be expressed in terms of universal constants of nature, such as Newton's gravitational constant G, Einstein's universal speed of light c, and Planck's constant h. The natural size of a universe is then tiny, about 10^{-33} cm, and the natural lifetime about 10^{-43} seconds, far from the sizes of our world. Physicists need to explain why our world is large and old and cold and dark. Quantum theory provides the opportunity to connect the Planck scales with our scales, our world, and our physical laws, because in quantum theory virtual particles of all masses enter the equations and mix scales.

But that only works if the underlying theory is what is called a supersymmetric one, with our familiar particles, such as quarks and electrons and force-mediating bosons each having a superpartner particle (squarks, selectrons, gluinos, etc.). In collisions at the LHC, the higher energy of the colliding particles turns into the masses of previously unknown particles via Einstein's $E=mc^2$.

The theory did not tell us how massive the superpartners should be. Naïvely, there were arguments that they should not be too heavy ("naturalness"), so they could be searched for with enthusiasm at every higher energy that became accessible, but so far they have not been found. In the past decade or so, string theory and M-theory have been better understood and now provide clues as to how heavy the superpartners should be. String theories and M-theories differ technically in ways not important for us here. To be mathematically consistent, and part of a quantum theory of gravity and the other forces, they must have nine or ten space dimensions (and one time dimension). To predict superpartner masses, they must be projected onto our world of three space dimensions, and there are known techniques for doing that.

The bottom line is that well-motivated string/M-theories do indeed predict that the LHC run (Run II), which started in late 2015 and is moving ahead strongly in 2016, should be able to produce and detect some superpartners, thus opening the door to the Planck-scale world and promoting study of a final theory to testable science. The news that the LHC works at its full energy and intensity and is expected to accumulate data for several years is a strong candidate for the most important scientific news of recent years.

NEW PROBES OF EINSTEIN'S CURVED SPACETIME—AND BEYOND?

STEVE GIDDINGS

Theoretical physicist; professor, Department of Physics, UC Santa Barbara

One of the most profound puzzles in modern physics is to describe the quantum nature of spacetime. A real challenge here is that of finding helpful experimental guidance. Interestingly, we are just now on the verge of gaining key new experimental information about classical spacetime, in new and important regimes—and this offers a possibility of learning about quantum spacetime as well.

The community has been abuzz about the possible discovery of a new particle at the LHC, seen by its disintegration into pairs of photons. If this is real and not just a fluctuation, there's a slim chance it is a graviton in extra dimensions, which, if true, could well be the discovery of the century. While this would indeed be a probe of quantum spacetime, I'll put it aside until more data reveals what's happening at the LHC.

But we are clearly entering a new era in several respects. First, miles-long instruments built to detect gravitational waves have just reached a sensitivity where they should be able to see these spacetime ripples, emitted from collisions and mergers of distant black holes and neutron stars. In fact, at this writing there have been recent hints of signals seen in these detectors, though we are awaiting a verifiable signal. Once found, these will confirm a major prediction of Einstein's general relativity and open a new branch of astronomy, where distant objects are studied by the gravitational waves they emit. It's also possible that precise measurements of the microwave radiation left over from the Big Bang will reveal gravity waves, though the community has backpedaled from the premature announcement of this in 2014, and so the race may well be won by the Earth-based gravity-wave detectors. Either way, developments will be exciting to watch.

Even more profound tests of general relativity may well exist via the Event Horizon Telescope, which is being brought online to study the 4-million-solar-mass black hole at the center of our galaxy. The EHT is really a network of radio telescopes, which combine to make a telescope the size of planet Earth. This will offer an unprecedentedly sharp focus on both our central black hole and on the 6-*billion*-solar-mass black hole at the center of the nearby elliptical galaxy M87. In fact, with the telescopes that have been networked so far, we're beginning to see structure whose size is close to that of the event horizon of our central black hole. The EHT should ultimately probe gravity in a regime where it gets extremely strong—so strong that the velocity needed to escape its pull approaches the speed of light. This will give us a new view on gravity in a regime where it has so far not been well tested.

Even more tantalizing is the possibility that the EHT will start to see effects that begin to reveal a more basic quantum reality underlying spacetime. For the 2014 *Edge* Question, I wrote that our fundamental concept of spacetime seems ready for retirement and needs to be replaced by a more basic quantum structure. There are many reasons for this: One good one is the crisis arising from the attempt to explain black-hole evolution with present-day physics; our current foundational principles, including the idea of spacetime, conflict with each other in describing black holes. Although Stephen Hawking initially predicted that quantum mechanics must break down when we account for emission of particles of Hawking radiation from a black hole, there are now good indications that it should not be abandoned. And if quantum mechanics is to be saved, this tells us that quantum information must be able to escape a black hole as it radiates particles—and this confronts our understanding of spacetime.

The need for information to escape an evaporating black hole conflicts with our current notions of how fields and particles move in spacetime; here, escape is forbidden by the prohibition of faster-than-light travel. A key question is how the familiar spacetime picture of a black hole must be modified to allow such escape. The modifications apparently must extend out at least to the hole's event horizon. Some have postulated that the new effects abruptly stop right there, at the horizon, but this abruptness is unnatural and leads to other seemingly crazy conclusions associated with what has been recently renamed the "firewall" scenario. A more natural scenario is that the usual spacetime description is also modified in a region extending outward beyond the black-hole horizon, at least through the region where gravity is very strong; the size of this region is perhaps a few times the horizon radius. In short, the need to save quantum mechanics indicates quantum modifications to our current spacetime description that extend into the region that the EHT observations will be probing! Important goals are to improve understanding of the nature of these alterations to the familiar spacetime picture and determine more carefully their possible observability via the EHT's measurements.

SUPERMASSIVE BLACK HOLES

JEREMY BERNSTEIN

Professor emeritus, Stevens Institute of Technology; former staff writer, *The New Yorker*

The most interesting thing I learned was the presence of supermassive black holes at the center of galaxies, including our own. Where did they come from? At what stage were they created? They are not the collapse of a star. I have not heard an explanation that makes a lot of sense to me. Maybe they are pre-Big Bang relics.

GIGANTIC BLACK HOLES AT THE CENTER OF GALAXIES

CARLO ROVELLI

Theoretical physicist, Centre de Physique Théorique, Aix-Marseille University, Marseille, France; author, *Seven Brief Lessons on Physics*

Evidence has recently piled up that there is a gigantic black hole, Sagittarius A★, with a mass 4 million times that of our Sun, at the center of our galaxy. Similar black holes appear to exist at the center of most galaxies. Some have masses *billions* of times that of our Sun. Can you imagine a black hole a billion times the size of our Sun?

The existence of these giants changes, once again, our picture of the universe. Clearly these monsters must have played a major role in the history of the cosmos but we do not know what it was. Astronomers are building an "instrument" as large as the Earth, connecting many existing radio telescopes to see Sagittarius A★ directly.

But these immense black holes are also the boundary of our current knowledge: We see matter falling into them, but we have no idea what ultimately happens to it. Space and time appear to come to an end inside. Or, better said, to morph into something we do not yet know. The universe is still full of mystery.

THE UNIVERSE IS INFINITE

RUDY RUCKER

Mathematician, computer scientist, cyberpunk pioneer; co-author
(with Bruce Sterling), *Transreal Cyberpunk*

Many cosmologists now think our spatial universe is infinite.
That's news. It was only this year that I heard about it. I don't
get out as much as I used to.

Thirty years ago, it was widely believed that our spatial
universe is the finite 3D hypersurface of a 4D hypersphere—
analogous to the finite 2D surface of a 3D sphere. Our underlying hypersphere was supposedly born, and began expanding, at
the Big Bang. And eventually our hypersphere was to run out of
momentum and collapse back into a Big Crunch—which might
possibly serve as the seed for a new Big Bang. No yawning void
of infinity and no real necessity for a troublesome initial point in
time. Our own Big Bang itself may have been seeded by a prior
Big Crunch. Indeed, we could imagine an endless pearl-string
of successive hyperspherical universes. A tidy theory.

But then experimental cosmologists found ways to estimate
the curvature of our space, and it seems to be flat, like an endless plane, not curved like the hypersurface of a hypersphere. At
most, our space might be "negatively curved," like a hyperbolic
saddle shape, but that's probably infinite as well.

If you're afraid of infinity, you might say something like this:
"So, OK, maybe we're in a vast infinite space, but it's mostly
empty. Our universe is just a finite number of galaxies rushing
away from each other inside this empty infinite space—like a

solitary skyrocket exploding and sending out a doomed shower of sparks." But many cosmologists say, "No, there are an infinite number of galaxies in our infinite space."

Where did all those galaxies come from? The merry cosmologists deploy a slick argument involving the relativity of simultaneity and the inflationary theory of cosmic inflation—and they conclude that in the past there was a Big Bang explosion at every single point of our infinite space. *Flaaash!* An infinite space with infinitely many galaxies!

Note that I'm not talking about some shoddy "many universes" theory here. I hate those things. I'm talking about our good old planets-and-suns single universe. And they're telling us it goes on forever in space, and on forever into the future, and it has infinitely many worlds. We aren't ever going to see more than a few of these planets, but it's nice to know they're out there.

So, OK, how does this affect me in the home?

You get a sense of psychic expansion if you begin thinking in terms of an infinite universe. A feeling of freedom, and perhaps a feeling that whatever we do here does not, ultimately, matter that much. You'd do best to take this in a "Relax!" kind of way, rather than in an "It's all pointless" kind of way.

Our infinite universe's inhabited planets are like dandelions in an endless meadow. Each of them is beautiful and to be cherished—especially by the little critters who live on them. We cherish our Earth because we're part of it, even though it's nothing special. It's like the way you might cherish your family. It's not unique, but it's yours. And maybe that's enough.

I know some of you are going to want more. Well, as far as I can see, we're living in one of those times when cosmologists have no clear idea of what's going on. They don't understand the start of the cosmos, nor cosmic inflation, nor dark energy, nor dark matter. You might say they don't know jack.

Not knowing jack is a good place to be, because it means we're ready to discover something really cool and different. Maybe next year, maybe in ten, or maybe in twenty years. Endless free energy? Antigravity? Teleportation? Who can say. The possibilities are infinite and the future is bright.

It's good to be an infinite world.

ADVANCED LIGO AND ADVANCED VIRGO

PAUL DAVIES

Theoretical physicist, cosmologist, astrobiologist, Arizona State University; author, *The Eerie Silence: Renewing Our Search for Alien Intelligence*

The end of 2015 coincided with the centennial of Einstein's general theory of relativity, which the great man presented to the Prussian Academy of Sciences in a series of four lectures in the midst of World War I. Widely regarded as the pinnacle of human intellectual achievement, general relativity took many years to be well tested observationally. But after decades of thorough investigation, physicists have yet to find any flaw with the theory.

Nevertheless, one key test remains incomplete. Shortly after Einstein published his famous gravitational field equations, he came up with an intriguing solution of them. It describes ripples in the geometry of spacetime itself, representing waves that travel across the universe at the speed of light. The detection of these gravitational waves has been an outstanding challenge to experimental physics for several decades. Now, in early 2016, the long search seems to be nearing its culmination.

A laser system designed to pick up the passage of gravitational waves emanating from violent astronomical events has recently been upgraded, and rumors abound that it has already seen something. The system, called Advanced LIGO (for Laser Interferometer Gravitational-wave Observatory), uses laser beams to spot almost inconceivably minute gravitational effects. In Europe, its counterpart, Advanced Virgo, is also limbering up. Advanced LIGO and Advanced Virgo are refinements of existing systems that proved the technology but lacked the sensitivity to detect bursts of gravitational waves from supernovae or colliding neutron stars on a routine basis. The stage is now set to move to that phase.

The detection of gravitational waves would not merely provide a definitive test of Einstein's century-old theory; it would serve to open up a whole new window on the universe. Existing conventional telescopes range across the entire electromagnetic spectrum, from radio to gamma rays. LIGO and Virgo would open up an entirely new spectrum and with it an entirely new branch of astronomy, enabling observations of black-hole collisions and other cosmic exotica.

Each time a new piece of technology has been used to study the universe, astronomers have been surprised. Once gravitational astronomy is finally born, the exploration of the universe through gravitational eyes will undoubtedly provide newsworthy discoveries for decades to come.

THE NEWS IS NOT THE NEWS

FRANK WILCZEK

Theoretical physicist, MIT; recipient, 2004 Nobel Prize in physics; author, *A Beautiful Question: Finding Nature's Deep Design*

On the ice-capped heights of Labrador, through winter, snow falls. With the coming of spring, much of it melts. Sometimes more falls than melts, and the ice grows; sometimes more melts than has fallen, and the ice shrinks. It is a delicate balance. The result varies from year to year, by many inches. But let the balance tip ever so slightly, so that amid much larger fluctuations one inch, on average, survives, and Earth is transformed. Great glaciers grow and cover North America in ice. If corresponding processes in Greenland or Antarctica tip the other way, melting more than is frozen, then oceans will swell and drown North America's coasts.

Episodes of both sorts have happened repeatedly in Earth's history, on timescales of a few tens of thousands of years. They are probably controlled by small, long-period changes in Earth's orbit. Today we are living in a relatively rare interglacial period, expected to last another 50,000 years. Notoriously, over the last few decades human activity has tipped the balance toward melting, threatening catastrophe.

These mighty stories derive from systematic trends that can be hard to discern within the tumult of much larger but ephemeral noise. The news is not the news.

So it is with the grandest of human stories: the steady increase, powered by science, of our ability to control the physical world. Richard Feynman memorably expressed a related thought: "From a long view of the history of mankind, seen from, say, ten thousand years from now, there can be little doubt that the most significant event of the 19th century will be judged as Maxwell's discovery of the laws of electrodynamics."

In that spirit, the most significant event of the 20th century is the discovery of the laws of matter in general. That discovery has three components: the frameworks of relativity and quantum mechanics, and the specific forces and laws embodied in our core theory, often called the Standard Model. For purposes of chemistry and engineering—plausibly, for all practical purposes—we've learned what nature has on offer.

I venture to guess that the most significant event of the 21st century will be a steady accumulation of new discoveries, based on deeper use of quantum physics, which harness the physical world. In the 21st century, we will learn how to harvest energy from the Sun and store it efficiently. We will learn how to make much stronger, much lighter materials. We will learn how to make more powerful and more versatile illuminators, sensors, communication devices, and computers.

We know the rules. Aided by our own creations, in a virtuous cycle, we'll learn how to play the game.

WE KNOW ALL THE PARTICLES AND FORCES WE'RE MADE OF

SEAN CARROLL

Theoretical physicist, Caltech; author, *The Big Picture: On the Origins of Life, Meaning, and the Universe Itself*

Sometimes news creeps up on us slowly. The discovery of the electron by J. J. Thomson in 1897 marked the first step in constructing the Standard Model of particle physics, an endeavor that culminated in the discovery of the Higgs boson in 2012. The Standard Model is a boring name for a breathtaking theory describing quarks, leptons, and the bosons that hold them all together to make material objects. Together with gravity, captured by Einstein's general theory of relativity, we have what Nobel Laureate Frank Wilczek has dubbed the core theory: a complete description of all the particles and forces that make up you and me, as well as the Sun, Moon, and stars, and everything we've directly seen in every experiment performed here on Earth.

There is a lot we don't understand in physics: the nature of dark matter and dark energy, what happens at the Big Bang or inside a black hole, why the particles and forces have the characteristics they do. We certainly don't know even a fraction of what there is to learn about how the elementary particles and forces come together to make complex structures, from molecules to nation-states. But there are some things we do know—and that includes the identity and behavior of all of the pieces underlying the world of our everyday experience.

Could there be particles and forces we haven't yet discovered? Of course—there almost certainly are. But the rules of quantum field theory assure us that if new particles and forces interacted strongly enough with the ones we know about to play any role in the behavior of the everyday world, we would have been able to produce them in experiments. We've looked, and they're not there. Any new particles must be too heavy to be created, or too short-lived to be detected; any new forces must be too short-range to be noticed, or too feeble to push around the particles we see. Particle physics is nowhere near complete, but future discoveries in that field won't play a role in understanding human beings or their environment.

We'll continue to push deeper. There's a very good chance that "particles and forces moving through spacetime" isn't the most fundamental way of thinking about the universe. Just as we realized in the 19th century that air and water are fluids made of atoms and molecules, we could discover that there is a layer of reality more comprehensive than anything we currently imagine. But air and water didn't stop being fluids just because we discovered atoms and molecules; we still give weather reports in terms of temperature and pressure and wind speed, not by listing what each individual molecule in the atmosphere is doing. Similarly, 1,000 and 1,000,000 years from now we'll still find the concepts of the core theory to be a useful way of talking about what we're made of.

Could we be wrong in thinking that the core theory describes all of the particles and forces that go into making human beings and their environments? Sure, we could always be wrong. The Sun might not rise tomorrow, we could be brains living in vats, or the universe could have been created last Thursday. Science is an empirical enterprise, and we should always be willing to change our minds when new evidence comes in. But quantum field theory is a special kind of framework. It's the unique way of accommodating the requirements of quantum mechanics, relativity, and locality. Finding that it was violated in our everyday world would be one of the most surprising discoveries in the history of science. It could happen—but the smart money is against it.

The discovery of the Higgs boson at the Large Hadron Collider in 2012 verified that the basic structure of the core theory is consistent and correct. It stands as one of the greatest accomplishments in human intellectual history. We know the basic building blocks of which we are made. Figuring out how those simple pieces work together to create our complex world will be the work of many generations to come.

COMPUTATIONAL COMPLEXITY AND THE NATURE OF REALITY

AMANDA GEFTER

Science writer; author, *Trespassing on Einstein's Lawn*

Physicists have spent the last 100 years attempting to reconcile Einstein's theory of general relativity, which describes the large-scale geometry of spacetime, with quantum mechanics, which deals with the small-scale behavior of particles. It's been slow going for a century, but now, suddenly, *things are happening*.

First, there's *ER=EPR*. More idea than equation, it's the brainchild of physicists Juan Maldacena and Leonard Susskind. On the lefthand side is an Einstein-Rosen bridge, a kind of geometric tunnel connecting distant points in space, otherwise known as a wormhole. On the right are Einstein, Rosen, and Podolsky, the three physicists who first pointed out the spooky nature of quantum entanglement, wherein the quantum state of two remote particles straddles the distance between them. Then there's that "equals" sign in the middle. Boldly it declares that spacetime geometry and the links between entangled particles are two descriptions of the same physical situation. *ER = EPR* appears brief and unassuming, but it's a daring step toward uniting general relativity and quantum mechanics—with radical implications.

Intuitively, the connection is clear. Both wormholes and entanglement flout the constraints of space. They're shortcuts. One can enter a wormhole on one side of the universe and emerge from it on the other without having to traverse the space in between. Likewise, measuring one particle will instantaneously determine the state of its entangled partner, even if the two are separated by galaxies.

The connection becomes more intriguing when viewed in terms of information. For maximally entangled particles, the information they carry resides simultaneously in both particles but in neither alone; informationally speaking, no space separates one from the other. For particles that are slightly less than maximally entangled, we might say there is some space between them. As particles become less and less entangled, information becomes more and more localized, words like "here" and "there" begin to apply and ordinary space emerges.

One hundred years ago, Einstein gave us a new way to think about space—not as the static backdrop of the world but as a dynamic ingredient. Now *ER=EPR* gives us yet a newer interpretation: What we call space is nothing more than a way to keep track of quantum information. And what about time? Time, physicists are beginning to suspect, may be a barometer of computational complexity.

Computational complexity measures how difficult it is to carry out a given computation—how many logical steps it takes, or how the resources needed to solve a problem scale with its size. Historically, it's not something physicists thought much about. Computational complexity was a matter of engineering— nothing profound. But all that has changed, thanks to what's known as the black-hole firewall paradox—an infuriating dilemma that has theoretical physicists pulling out their hair.

As a black hole radiates away its mass, all the information that ever fell in must emanate back out into the universe, scrambled among the Hawking radiation; if it doesn't, quantum mechanics is violated. The very same information must reside deep in the black hole's interior; if it doesn't, general relativity is violated. And the laws of physics decree that information can't be duplicated. The firewall paradox arises when we consider an observer, Alice, who decodes the information scrambled among the Hawking radiation, then jumps into the black hole where she will find, by various accounts, an illegal information clone or an inexplicable wall of fire. Either way, it's not good.

But Alice's fate recently took a turn when two physicists, Patrick Hayden and Daniel Harlow, wondered how long it would take her to decode the information in the radiation. Applying a computational-complexity analysis, they discovered that the decoding time would rise exponentially with each additional particle of radiation. In other words, by the time Alice decodes the information, the black hole will have long ago evaporated and vanished, taking any firewalls or violations of physics with it. Computational complexity allows general relativity and quantum mechanics to peacefully coexist.

Hayden and Harlow's work connects physics and computer science in a totally unprecedented way. Physicists have long speculated that information plays a fundamental role in physics. It's an idea that dates back to Konrad Zuse, the German engineer who built the world's first programmable electronic computer in his parent's living room in 1938 and the first universal Turing machine three years later. In 1969, Zuse wrote a book called *Calculating Space*, in which he argued that the universe itself is a giant digital computer. In the 1970s, the physicist John Wheeler began advocating for "it from bit"—the notion that the physical, material world is, at bottom, made of information. Wheeler's influence drove the burgeoning field of quantum information theory and led to quantum computing, cryptography, and teleportation. But the idea that computational complexity might not only describe the laws of physics but actually *uphold* the laws of physics is entirely new.

It's odd, on first glance, that something as practical as resource constraints could tell us anything deep about the nature of reality. And yet in quantum mechanics and relativity, such seemingly practical issues turn out to be equally fundamental. Einstein deduced the nature of spacetime by placing constraints on what an observer can see. Noticing that we can't measure simultaneity at a distance gave him the theory of special relativity; realizing that we can't tell the difference between acceleration and gravity gave him the curvature of spacetime. Likewise, when the founders of quantum mechanics realized that it is impossible to accurately measure position and momentum, or time and energy, simultaneously, the strange features of the quantum world came to light. That such constraints were at the heart of both theories led thinkers such as Arthur Stanley Eddington to suggest that at its deepest roots physics is epistemology. The new computational complexity results push further in that direction.

So that's the news: a profound connection between information, computational complexity, and spacetime geometry has been uncovered. It's early to say where these clues will lead, but it's clear now that physicists, computer scientists, and philosophers will all bring something to bear to illuminate the hidden nature of reality.

EINSTEIN WAS WRONG

HANS HALVORSON
Professor of philosophy, Princeton University

We've known about "quantum weirdness" for more than 100 years, but it's still making headlines. In the summer of 2015, experimental groups in Boulder, Delft, and Vienna announced that they had completed a decades-long quest to demonstrate quantum nonlocality. The possibility of such nonlocal effects first captured the attention of physicists in the 1930s, when Einstein called it "spooky action at a distance"—indicating that he perceived it as a bug of the nascent theory. But on this particular issue, Einstein couldn't have been more wrong: Nonlocality isn't a bug of quantum mechanics, it's a pervasive feature of the physical world.

To understand why the scientific community has been slow to embrace quantum nonlocality, recall that 19th-century physics was built around the ideal of *local causality*. According to this ideal, for one event to cause another, those two events must be connected by a chain of spatially contiguous events. In other words, for one thing to affect another, the first thing needs to touch something, which touches something else, which touches something else . . . eventually touching the other thing.

For those of us schooled in classical physics, the notion of local causality might seem central to a rational outlook on the physical world. For example, I don't take reports of telekinesis seriously—and not because I've taken the time to examine all the experiments that have tried to confirm its existence. No, I don't take reports of telekinesis seriously because it seems irrational to believe in some sort of causality that doesn't involve things moving through space and time.

But QM appears to conflict with local causality. According to QM, if two particles are in an entangled state, then the outcomes of measurements on the second particle will always be strictly correlated (or anticorrelated) with measurements on the first particle—even when the second particle is far, far away from the first. Quantum mechanics also claims that neither the first nor the second particle has any definite state before the measurements are performed. So what explains the correlations between the measurement outcomes?

It's tempting to think that quantum mechanics is just wrong when it says that the particles aren't in any definite state before they're measured. In fact, that's exactly what Einstein suggested in the famous "*EPR*" paper he wrote with Podolsky and Rosen. However, in the 1960s, John Stewart Bell showed that *EPR* could be put to experimental test. If, as suggested by Einstein, each particle has its own state, then the results of a certain crucial experiment would disagree with the predictions made by quantum mechanics. Thus, in the 1970s and 1980s, the race was on to perform this crucial experiment—an experiment that would establish the existence of quantum nonlocality.

The experiments of the 1970s and 80s came out decisively in favor of quantum nonlocality. However, they left open a couple of loopholes. It was only in 2015 that the ingenious experimenters in Boulder, Delft, and Vienna were able to definitively close these loopholes—propelling quantum nonlocality back into the headlines.

But is it news that quantum mechanics is true? Didn't we already know this—or at least wasn't the presumption strongly in its favor? Yes, the real news here isn't that quantum mechanics is true. The real news is that we are learning how to harness the resources of a quantum world. In the 1920s and 1930s, quantum nonlocality was a subject of philosophical perplexity and debate. In 2015, questions about its meaning are being replaced by questions about what we can do with it. For instance, quantum nonlocality could facilitate information-theoretic and crytographic protocols that far exceed anything that could have been imagined in a world governed by classical physics. And this is the reason quantum nonlocality is still making headlines.

But don't get carried away—quantum nonlocality still doesn't make it rational to believe in telekinesis.

REPLACING MAGIC WITH MECHANISM?

ROSS ANDERSON

Professor of security engineering, Computer Laboratory, University of Cambridge; author, *Security Engineering*

The most thought-provoking scientific meeting I went to in 2015 was Emergent Quantum Mechanics, organized in Vienna by Gerhard Groessing. This is the go-to place if you're interested in whether quantum mechanics dooms us to a universe (or multiverse) that can be causal or local but not both, or whether we might just make sense of it after all. The big new theme was emergent global correlation. What is this, and why does it matter?

The core problem of quantum foundations is the Bell tests. In 1935, Einstein, Podolsky, and Rosen noted that if you measured one of a pair of particles that shared the same quantum-mechanical wave function, this would immediately affect what could be measured about the other even if it were some distance away. Einstein held that this "spooky action at a distance" was ridiculous so quantum mechanics must be incomplete. This was the most cited paper in physics for decades. In 1964, the Irish physicist John Bell proved that if particle behavior were explained by hidden local variables, their effects would have to satisfy an inequality that would be broken in some circumstances by quantum-mechanical behavior. In 1969, Clauser, Horne, Shimony, and Holt proved a related theorem that limits the correlation between the polarization of two photons, assuming that this polarization is carried entirely by and within them. In 1974,

Freedman and Clauser showed this limit was violated experimentally, followed by Aspect, Zeilinger, and many others. These "Bell tests" convince many physicists that reality must be weird; maybe nonlocal, noncausal, or even involving multiple universes.

For example, it's possible to entangle photon A with photon B, then B with C, then C with D, and measure that A and D are correlated, despite the fact that they didn't exist at the same time. Does this mean that when I measure D, some mysterious influence reaches backward in time to A? The math doesn't let me use this to send a message backward in time—say, to order the murder of my great-grandfather (the no-signaling theorem becomes a kind of "no-TARDIS theorem")—but such experiments are still startlingly counterintuitive.

At the Vienna conference, a number of people advanced models according to which quantum phenomena emerge from a combination of local action and global correlation. As the Nobel prizewinner Gerard 't Hooft put it in his keynote talk, Bell assumed that spacelike correlations are insignificant, and this isn't necessarily so. In Gerard's model, reality is information, processed by a cellular automaton fabric operating at the Planck scale, and fundamental particles are virtual particles—like Conway's gliders but in three dimensions. In a version he presented at the previous EmQM event in 2013, the fabric is regular, and its existence many break gauge invariance just enough to provide the needed long-range correlation. The problem was that the Lorentz group is open, which seemed to prevent the variables in the automata being bitstrings of finite length. In his new version, the automata are randomly distributed. This was inspired by an idea of Stephen Hawking's on balancing the information flows into and out of black holes.

In a second class of emergence models, the long-range order comes from an underlying thermodynamics. Groessing has a model in which long-range order emerges from subquantum statistical physics; Ariel Caticha has a model with a similar flavor, which derives quantum mechanics as entropic dynamics. Ana María Cetto looks to the zero-point field and sets out to characterize active zero-point field modes that sustain entangled states. Bei-Lok Hu adds a stochastic term to semiclassical gravity, whose effect after renormalization is nonlocal dissipation with colored noise.

There are others. The quantum-cryptography pioneer Nicolas Gisin has a new book on quantum chance in which he suggests that the solution might be nonlocal randomness—a random event that can manifest itself at several locations. My own suspicion is that it might be something less colorful; perhaps the quantum vacuum just has an order parameter, like a normal superfluid or superconductor. If you want long-range order that interacts with quantum systems, we have quite a few examples and analogs to play with.

But whether you think the quantum vacuum is God's computer, God's bubble bath, or even God's cryptographic keystream generator, there's suddenly a sense of excitement and progress, of ideas coming together, of the prospect that we might just possibly be able to replace magic with mechanism.

There may be a precedent. For about forty years after Galileo, physics was a free-for-all. The old Ptolemaic certainties had been shot away and philosophers' imaginations ran wild. Perhaps it would be possible, some said, to fly to America in eight hours in a basket carried by swans? Eventually Newton wrote the *Principia* and spoiled all the fun. Theoretical physics has been stuck for the past forty years, and imaginations have been running wild once more. Multiple universes that let stuff travel backward in time without causing a paradox? Or perhaps it's time for something new to come along and spoil the fun.

QUANTUM ENTANGLEMENT IS INDEPENDENT OF SPACE AND TIME

ANTON ZEILINGER

Physicist, University of Vienna; scientific director, Institute of Quantum Optics and Quantum Information; author, *Dance of the Photons*

The notion of quantum entanglement, famously called "spooky action at a distance" by Einstein, emerges more and more as having deep implications for our understanding of the world. Recent experiments have verified the fact that quantum correlations between two entangled particles are stronger than any classical, local pre-quantum worldview allows. So, since quantum physics has predicted these measurement results for at least eighty years, what's the deal?

The point is that the predictions of quantum mechanics are independent of the relative arrangement in space and time of the individual measurements: fully independent of their distance, independent of which is earlier or later, etc. One has perfect correlations between all of an entangled system, even as these correlations cannot be explained by properties carried by the system before measurement. So quantum mechanics transgresses space and time in a very deep sense. We would be well advised to reconsider the foundations of space and time in a conceptual way.

To be specific, consider an entangled ensemble of systems. This could be two photons, or any number of photons, electrons, atoms—and even larger systems, like atomic clouds at low temperature or superconducting circuits. We now do measurements individually on those systems. The important point is that for a maximally entangled state, quantum physics predicts random results for the individual entangled property. For photons, say, this could the polarization. That is, for a maximally entangled state of two or more entangled photons, the polarization observed in the experiment could be anything: horizontal, vertical, any direction linear, right-handed circular, left-handed circular, any elliptical state. Thus, if we do a measurement, we observe a random polarization. And this for each individual photon of the entangled ensemble. But a maximally entangled state predicts perfect correlations between the polarizations of all photons making up the entangled state.

To me, the most important message is that the correlations between particles, like photons, electrons, or atoms—or larger systems, like superconducting circuits—are independent of which of the systems are measured first and how large the spatial distance between them is.

At first glance, this might not seem surprising. After all, if I measure the heights of peaks of the mountains around me, it doesn't matter in which sequence I do the measurements and whether I measure the more distant ones first or the nearer ones. The same is true for measurements on entangled quantum systems. However, the important point is that the first measurement on any system entangled with others instantly changes the common quantum state describing all, the subsequent measurement on the next does that again, and so on. Until, in the end, all measurement results on all systems entangled with each other are perfectly correlated.

Moreover, as recent experiments finally prove, we now know that this cannot be explained by any communication limited by Einstein's cosmic speed limit, the speed of light. Also, one might think that there is a difference if two measurements are done such that one is after the other in way that a signal could tell the second one what to do as a consequence of the earlier measurement. Or whether they are arranged at such a distance and done sufficiently simultaneously such that no signal is fast enough to do so. Thus, it appears that on the level of measurements of properties of members of an entangled ensemble, quantum physics is oblivious to space and time.

An understanding is possible via the notion of information—information seen as the possibility of obtaining knowledge. Then quantum entanglement describes a situation where information exists about possible correlations between possible future results of possible future measurements without any information existing for the individual measurements. The latter explains quantum randomness, the first quantum entanglement. And both have significant consequences for our customary notions of causality.

It remains to be seen what the consequences are for our notions of space and time—or spacetime, for that matter. Spacetime itself cannot be above or beyond such considerations. I suggest we need a new deep analysis of spacetime, a conceptual analysis perhaps analogous to the one done by the Viennese physicist-philosopher Ernst Mach, who kicked Newton's absolute space and absolute time from their throne. The hope is that in the end we will have new physics analogous to Einstein's new physics in the two theories of relativity.

BREAKTHROUGHS BECOME PART OF THE CULTURE

LISA RANDALL

Theoretical physicist, Harvard University; author, *Dark Matter and the Dinosaurs: The Astounding Interconnectedness of the Universe*

Some of the interesting discoveries and observations of the last year include a new species of human; observations of dwarf planets including Pluto, which demonstrated that Pluto was more geologically active than anticipated; data of species loss indicating a track toward a sixth extinction; and a careful measurement of the timing of the impact triggering the K-Pg extinction and enhanced Deccan Traps volcanic activity, indicating that they occurred at essentially the same time, thus both contributing to species loss 66 million years ago. But news in science is usually the product of many years of effort, even when it appears to be a sudden revolutionary discovery, and the headlines of any given year are not necessarily representative of what is most significant.

So I'm going to answer a slightly different question, which is what advances I expect we'll hear about in the coming decade, bearing in mind that the most common stories concern news that in some global sense hasn't changed all that much. Crisp clean events and many important discoveries are news, but for only a short time. True breakthroughs become part of the culture. General relativity was news in 1915 and the bending of light was news in 1919. Yet although general relativity factors into news today, the theory itself is no longer news. Quantum mechanics stays in the news, but only because people don't want to believe it, so incremental verifications are treated as newsworthy.

Instead of saying more about the important discoveries of the last year, I'll give a few examples of scientific advances I expect we might hear about in the next few. The first is the type we won't really solve but will marginally, incrementally develop. The second is a type where we will make advances but the news won't necessarily reflect the most important implications. The third might be a true breakthrough that largely solves a puzzle—like the Higgs boson discovery, which was big news in 2012 but (though still exciting and an important guidepost for the future of particle physics) is no longer news today.

The first discovery concerns a better understanding about what constitutes life—or at least life as we know it. We'll learn more about the chemical composition of stuff in our solar system and perhaps where the elements of life as we know it arose. We might learn more about the chemistry, or at least some physical properties, of planets in other solar systems, and perhaps deduce more about where life (if not necessarily complex life) might arise. We will probably also examine the fossil record in greater detail as new chemical and physical methods allow us to probe deeper into Earth's history. All this will stay news, since we won't know how life arose for a long time to come, but small pieces of the puzzle will continue to emerge.

There will also be many new developments in artificial intelligence and robotics. These advances fall into the second category, since a lot of the real news about the role of automation will occur behind the scenes, where technology will make some tasks we already do simpler or more effective, or will replace workers and reduce (or at least change the nature of) employment. We'll read about drones and medical robotics and advances in AI, but those factory robots won't be big news except for a few days on the business pages and of course for the families of workers who find themselves on unemployment lines.

I hope that the third category will include discoveries that tell us more about the fundamental nature of dark matter. Dark matter is the matter that carries five times the energy of ordinary matter and interacts with gravity but very little or not at all with light. Experiments already in the news look for dark matter in different ways. Some of these, like XENON1T and LUX-ZEPLIN, use huge containers of material deep underground which might detect a tiny recoil of a dark-matter particle passing through. Also possible is that dark matter annihilates when two dark-matter particles get together and turns into photons.

There are less conventional searches that might tell us more about the nature of dark matter and that rely on comparing simulations of how structures like galaxies form from dark-matter collapse with actual data exploring the distribution of stars or other matter in galaxies. These more detailed observations of the role of dark matter might reveal some interesting aspects of how it interacts. Perhaps dark matter has interactions or forces that familiar matter doesn't experience—just as dark matter doesn't experience forces like the electromagnetism of the visible world.

If such properties of dark matter are found, or if a dark-matter particle is discovered, scientists will try to learn more about its properties and the implications for cosmology and astrophysics. But that will be a long slog, from the perspective of outsiders. The true sign that dark-matter searches have succeeded will be that the discovery will be taken for granted and cease to be news.

SPACE EXPLORATION, NEW AND OLD

ROBERT PROVINE

Psychologist; research professor/professor emeritus, University of Maryland, Baltimore County; author, *Curious Behavior: Yawning, Laughing, Hiccupping, and Beyond*

Surprising images of dwarf planet Pluto from a flyby of the *New Horizons* spacecraft put space exploration back in the headlines as one of the biggest science stories of 2015. Instead of a barren, frigid globe, Plato proved to be colorful, contrasty, and complex, with diverse geological structure including mountains, valleys, and plains of water-ice and nitrogen-ice, evidence of past and present glacial flows, possible volcanoes spewing water-ice from a warmer core, and a thin atmosphere that extends hundreds of miles above the planetary surface. Similar insights are being gleaned from Charon, the largest of Pluto's five moons.

2015 also brought Pluto news of a more arcane sort. The historic 24-inch Clark refracting telescope of Lowell Observatory was refurbished and opened to the public for viewing. As a young Lowell employee, Clyde Tombaugh discovered Pluto in 1930 on photographic plates taken with a Lowell instrument. Telescopes, whether optical leviathans or the modest backyard variety, are spaceships for the eye and mind which provide a compelling sensory immediacy lacking in the pricier technological *tour de force* of a spacecraft. Recall the aesthetic impact of the starry night viewed from a dark country path, or seeing Saturn for the first time through a telescope. Although modern

Earth-based telescopes continue to provide astronomical break-throughs, old telescopes and the observatories that house them survive as domed, verdigris-covered cathedrals of science. The Pluto flyby of 2015 is an occasion to celebrate space exploration new and old, and the value of looking upward and outward.

PLUTO IS A BUMP IN THE ROAD

NICHOLAS A. CHRISTAKIS

Physician, social scientist; director, Human Nature Lab, Yale University; co-author (with James H. Fowler), *Connected*

On July 14 of last year, NASA's *New Horizons* spacecraft flew within 7,800 miles of Pluto, after traversing 3 billion miles since its 2006 launch, and began sending back astonishing and detailed images of mountains and plains composed of ice from the last planet in our system to be explored. But for me, these images were not the most newsworthy aspect of its mission.

The science involved in accomplishing this feat is amazing. *New Horizons* is an engineering marvel, with radioisotope power generation, sophisticated batteries, optical and plasma scientific instruments, complex navigation and telemetry, and so on. Its primary missions—all successfully completed—were to map the surface of Pluto and its main moon, Charon; to characterize the geology and composition of these bodies; and to analyze their atmosphere. In the process, *New Horizons* has also shed light on the formation of our solar system.

We succeed in this kind of solar-system exploration so reliably nowadays that it seems to us routine, just a bump in the road of our endless inquiry. But the exploration of Pluto is, alas, a bump in the road in another, rather more dispiriting way.

Most Americans (58 percent, in a 2011 Pew survey) are supportive of space exploration, valuing its contributions both to science and to national pride. And most Americans (59 percent, in a 2015 Pew survey) approve of sending astronauts into space. Yet Americans and their politicians appear unwilling to spend more money on our space program; a 2014 General Social Survey indicated that just 23 percent of Americans think we should do so. By contrast, 70 percent of Americans think we should spend more on education and 57 percent think we should spend more on health. NASA's budget has been roughly constant in real dollars since 1985 (it's now 0.5 percent of the federal budget), but it is well below its peak in 1965, when it was more than 4 percent of the federal budget (and it's below even the 1-percent level of 1990).

The captivating photos of Pluto occupied a week or so in the news cycle in the middle of last summer, and we moved on. What amazed me about the news from Pluto was that there weren't more people who found this accomplishment aston-ishing, that there was not even a more sustained support for space exploration. NASA's entirely sensible push to make both manned and unmanned space exploration reliable, standardized, and safe has had the unfortunate side effect of making it routine and even boring for many people. For me, this is the real news-worthy part of the Pluto mission.

My paternal grandfather, who was born in Greece in the 1890s, used to tell me that he simply could not believe that he had heard about the first heavier-than-air flight of the Wright brothers when it happened, in 1903, and had also watched the Moon landing in 1969. He had fought in World War I and would tell me stories about how his unit was transferred from Ankara to Kiev on horseback "when the Bolsheviks revolted" and about how, during World War II, he kept his family alive in Athens "when the Nazis invaded." But space exploration interested him more, because it was so much more optimistic. Humans had gone from skimming over a beach in a plane made of canvas and bicycle parts to operating a lunar lander in sixty-six years. It amazed him, and the pace and sheer wonder of it astonish me even as I write this.

I realize, of course, that the great accomplishments in space exploration of the 1960s and 70s were largely motivated by the Cold War. I realize as well that many people are now arguing that private enterprise should take over space exploration. And I know that commitment to space exploration is low because many see better uses for our money. Is it better to vaccinate children, care for the poor, and invest in public health and medical research rather than invest in space exploration? Part of my response is the customary one that science and discovery are the ultimate drivers of our wealth and security. But my main response is that this is a false dichotomy. The real question is whether we would rather wage war or colonize Mars. Which would be, and should be, more newsworthy? In this I think my grandfather had it right.

PLUTO NOW, THEN ON TO 550 AU

GREGORY BENFORD

Novelist and emeritus professor of physics and astronomy, UC Irvine; co-author (with Larry Niven), *Shipstar*

The most long-range portentous event of 2015 was NASA's *New Horizons* spacecraft arrowing by Pluto, snapping clean views of the planet and its waltzing moon system. It carries an ounce of Clyde Tombaugh's ashes, commemorating his discovery of Pluto in 1930. Tombaugh would have loved seeing the colorful contrasts of this remarkable globe, far out into the dark of near-interstellar space. Pluto is now a sharply seen world, with much to teach us.

As the spacecraft zooms near an iceteroid on New Year's Day, 2019, it will show us the first member of the chilly realm beyond, where primordial objects quite different from the wildly eccentric Pluto also dwell. These will show us what sort of matter made up the early disk that clumped into planets like ours—a sort of family tree of worlds. But that's just an appetizer. *New Horizons* is important not just for completing our first look at every major world in the solar system. It points outward, to a great theater in the sky, where the worlds of the galaxy itself are on display.

Beyond Pluto looms a zone where the Sun's mass acts as a giant lens, its gravitation focusing the light of other stars to a small area. Think of it as gravity gathering starlight into an intense pencil, focused down as dots on a chilly sphere. Einstein calculated such gravitational bending of light in 1912, though

Newton knew the effect should occur in his own theory of mechanics and optics.

Images of whole galaxies made by this effect were not discovered until 1988. Such magnification of light from a star and the planets near it naturally creates a telescope of unparalleled power. It can amplify images by factors that can vary from 100 million to a quadrillion, depending on frequency. This suggests using such power to study worlds far across the interstellar reaches. We have already detected more than 2,000 planets around other stars, thanks to the *Kepler* mission and other telescopes. We can sense the atmospheres of some, when they pass across our view of their stars, silhouetted against that glare. Many more will come.

The space telescopes envisioned for the next several decades can tease out information about a planet at interstellar distances only by studying how light reflects or absorbs changes. At best, such worlds will be dots of faint light. But at the lensing distance, under enormously better resolution, we can see the worlds themselves—their atmospheres and moons, their seas and lands, perhaps even their cities.

Hardy *New Horizons* now zooms along at about 15 km a second, or (more usefully said) at about 3 astronomical units (AU, the distance between Earth and our Sun) in a year, relative to the Sun. The focus spot of the Sun is 550 AU out, as Einstein predicted in 1936. *New Horizons* will take 180 years to get to that focus—and will be long dead, as its nuclear power supply fades. So future missions to put a telescope out there demand speeds ten or more times faster. (*Voyager*, flying after thirty-eight-plus years, is only 108 AU away from Earth.)

We know of ways to propel spacecraft to such speeds. Most involve flying near the Sun and picking up velocity by firing rockets near it, or getting a boost from its intense light using unfurled solar sails, and other astro-tricks. Those feats we can fashion within decades, if we wish.

Our goal could be to put an observing spacecraft that can maneuver out at the focus of "God's zoom lens"—a 70-billion-mile-long telescope that light takes more than three days to traverse. An observing spacecraft could see whatever is behind the Sun from it, many light-years away.

This would vastly improve our survey of other worlds, to pick off strings of stars and examine their planets. Using the Sun as a lens works on all wavelengths, so we could look for signs of life—say, oxygen in an atmosphere—and perhaps even eavesdrop on aliens' radio stations squawking into the galactic night. At first, such a telescope could scrutinize Alpha Centauri's planets, if it has some—the next big step before trying to travel there. The craft could trace out a spiral pattern perpendicular to its outward path, slightly shifting its position relative to scan the Alpha Centauri system. Then look farther still—because the focus effect remains beyond 550 AU as a spacecraft moves outward, still seeing the immense magnifications.

New Horizons may be the best-named spacecraft of all, for it does indeed portend fresh, bold perspectives.

THE UNIVERSE SURPRISED US, CLOSE TO HOME

LAWRENCE M. KRAUSS
Physicist, cosmologist, Arizona State University;
author, *A Universe from Nothing*

When the first close-up pictures of Pluto came in from the *New Horizon* satellite, which flew by the planet last year, they shocked pretty well everyone who had thought about the now-demoted dwarf planet. Common sense suggested that Pluto should be a frozen ball, with a pockmarked surface reflecting billions of years of comet impacts. Instead, what was revealed was a dynamic object, with mountains 3–4 km high and a huge, 1,000-km-wide plain of ice with no impact craters—which means this plain cannot be older than 100 million years. Which in turn implies that the surface of Pluto is dynamic. Since there are no other large planets nearby that might be sources of tidal heating, this means that Pluto still has an active internal engine continuing to mold its surface. We have no idea how that could be the case.

Similar surprises have accompanied flybys of other solar-system objects—namely, the subsurface liquid-water ocean and organic-water geysers of Saturn's moon Encedalus and the volcanoes on Jupiter's moon Io. While these oddities are now understood to be powered by the huge tidal influence of their giant host planets, no one had expected this kind of extreme activity.

As we peer out farther to other stars, we have found them to be rife with planetary systems once thought to be impossible:

gas giants like Jupiter and Saturn orbiting closer to their stars than Mercury is to our Sun. It had been thought that inner planets would be small and rocky and outer planets larger and gaseous, as in our own solar system; we now understand that dynamical effects may have caused large planets to migrate inward over time in these systems.

Similarly, classical dynamics had suggested that binary star systems should not contain planets, as gravitational perturbations would expel such orbiting objects. But planets have now been discovered around binary stars, suggesting some new stabilizing mechanism at work.

We are accustomed to recognizing that at the extremes of scale the universe is a mysterious place. For example, dark energy—the energy of empty space—appears to dominate the dynamics of the universe on its largest scales, producing a gravitational repulsion that is causing the expansion of the universe to accelerate. On small scales, we currently have no idea why the newly discovered Higgs particle is as light as it is, one of the reasons the four forces in nature have the vastly different strengths we measure on laboratory scales.

But what we are learning as we explore our solar neighborhood is that the physics governing the formation and evolution of planetary-scale objects like Pluto, Io, and Enceladus—physics we thought was well understood—is far richer and more complex than we imagined. This not only gives the lie to claims that there would be no more new results of relevance to understanding human-scale physics but also puts in perspective the hyperbolic claims that a quantum theory of gravity (such as the most popular candidates, superstrings or M-theory) would be a Theory of Everything. Although such a theory would be of vital importance for understanding the origin of the universe and the nature of space and time, it would be irrelevant for understanding complex phenomena on human scales, like the boiling of oatmeal or formations of sand on the beach.

While oatmeal and sand may not capture the public's imagination, the exotic new worlds inside and outside our solar system certainly do. And our recent discoveries suggest that much conventional wisdom about even our nearest neighbors, and physics as classical as Newton's, will have to be rethought. The result of such revisions will likely shed new light on vital questions, including the big one: Are we alone in the universe? It's hard to see how our cosmic backyard could get more interesting!

PROGRESS IN ROCKETRY

GEORGE DYSON

Science historian; author, *Turing's Cathedral: The Origins of the Digital Universe*

Toward the end of 2015, in close succession, two rockets left the ground, crossed the Kármán line (at 100 km altitude) into space, and returned intact under their own power to a soft landing on the surface of the Earth. In the space business, new rockets are launched at regular intervals, but the launch of a used rocket is important news.

In December 1966, Project Orion pioneer Theodore B. Taylor complained that the high cost of sending anything into even low Earth orbit was "roughly equivalent to using jet transport planes to carry freight from, let us say, Madrid to Moscow, making one flight every few weeks, throwing away each aircraft after each flight, and including the entire construction and operation costs of several major airports in the cost of the flights!"

The now-abandoned space shuttle was a reusable spacecraft but failed to reduce launch costs and violated one of the cardinal rules of transport: Separate the passengers from the freight. Someday we will look back and recognize one of the other roadblocks to an efficient launch system: separating the propellant from the fuel.

There is no reason the source of reaction mass (propellant) has to be the same as the source of energy (fuel). Burning a near-explosive mix of chemicals makes the process inherently dangerous and places a hard limit on specific impulse (ISP), a measure of how much acceleration can be derived from a given amount of propellant/fuel. It is also the reason that the original objective of military rocketry—"to make the target more dangerous than the launch site"—took so long to achieve.

The launch business has been crippled, so far, by a vicious circle that has limited the market to expensive payloads— astronauts, military satellites, communication satellites, deep space probes—consigned by customers who can afford to throw the launch vehicle away after a single use. Reusable rockets are the best hope of breaking this cycle and moving forward on a path leading to low-cost, high-duty-cycle launch systems where the vehicle carries inert propellant and the energy source remains on the ground.

All the advances in autonomous control, combustion engineering, and computational fluid dynamics that allowed those two rockets to make a controlled descent after only a handful of attempts are exactly what will be needed to develop a new generation of launch vehicles that leave chemical combustion behind to ascend on a pulsed energy beam.

We took an important first step in this direction in 2015.

THE SPACE AGE TAKES OFF . . . AND RETURNS TO EARTH AGAIN

PETER SCHWARTZ

Futurist, business strategist; senior vice president for global
government relations and strategic planning, Salesforce.com;
author, *Inevitable Surprises*

As an adolescent in the fifties, along with many others, I dreamt
of the Space Age. We knew what the Space Age was supposed
to look like: silver, bullet-shaped rockets rising into the sky on a
column of flame, and as they return, descending on an identical
column of flame, landing gently at the spaceport. Those dreams
led me to a degree in astronautical engineering at RPI.

The reality of spaceflight turned out to be very different.
We built multistage booster rockets that were thrown away after
every launch. Bringing them back turned out to be too hard.
Carrying enough fuel to power the landing, and managing
the turbulent flow of the rocket exhaust as the vehicle slowly
descends on that violent, roaring column of flaming gas, was
too great a challenge. Indeed, even the efforts to build vertical
takeoff-and-landing jet fighters in the fifties also failed, for similar reasons.

The disposable launch vehicle made the Space Age too
costly for most applications. Getting any mass into orbit costs
many thousands of dollars per pound. Imagine what an airline
ticket would cost if the airline threw away the aircraft after every
flight. The booster vehicle generally costs a few hundred million
dollars—about the cost of a modern jet liner, and we get only

one use out of it. No other country, like Russia or China, nor any companies, like Boeing or Lockheed, could solve the technical problems of a reusable booster.

The space shuttle was intended to meet this challenge by being reusable. Unfortunately, the cost of refurbishing it after each launch was so great that the shuttle launch was far more expensive than a disposable launcher flight. When I worked on mission planning for the space shuttle at the Stanford Research Institute in the early seventies, the assumption was that the cost of each launch would be $118 per pound ($657 in current dollars), justifying many applications, with each shuttle flying once a month. Instead, the shuttles could fly only a couple of times per year, at a cost of $27,000 per pound, meaning most applications were off the table. So space was inaccessible except for those whose needs justified the huge costs either of a shuttle or a single-use booster: the military, telecommunications companies, and some government-funded high-cost science.

But in the last few weeks of 2015, all that changed, as the teams from two startup rocketry companies, Blue Origin and SpaceX, brought their launchers back to a vertical landing at the launch site. Both of their rockets were able to control that torrent of flaming gas to produce a gentle landing, ready to be prepared for another launch. Provided that we can do this on a regular basis, the economics of spaceflight have suddenly and fundamentally changed. It won't be cheap yet, but many more applications will be possible. And the costs will continue to fall with experience.

While both companies solved the hard problem of controlling the vehicle at slow speed on a column of turbulent gas, the SpaceX achievement will be more consequential in the near future. The Blue Origin rocket could fly to an altitude of only

60 miles before returning to Earth and is intended mainly for tourism. The SpaceX vehicle, *Falcon 9*, could (and did) launch a second stage that achieved Earth orbit. And SpaceX already ferries supplies, and may soon be carrying astronauts, to the International Space Station. The ability to reuse their most expensive component will reduce their launch costs by as much as 90 percent, and over time those costs will decline. Boeing and Lockheed should be worried.

Of course, the Blue Origin rocket, *New Shepard*, will also continue to improve. Their real competition is with Virgin Galactic, which has had some difficulties lately—a crash that killed one of the pilots. Both companies are competing for the space tourism market and (for now) Blue Origin appears to be ahead.

We have turned a corner in spaceflight. We can dream of a Space Age again. Life in orbit becomes imaginable. Capturing asteroids to mine and human interplanetary exploration have both become much more likely. The idea that many of us living today will be able to see Earth from space is no longer a distant dream.

HOW WIDELY SHOULD WE DRAW THE CIRCLE?

SCOTT AARONSON

David J. Bruton Centennial Professor of computer science, University of Texas at Austin; author, *Quantum Computing Since Democritus*

For fifteen years, popular-science readers have got used to breathless claims about commercial quantum computers being just around the corner. As far as I can tell, 2015 marked a turning point. For the first time, the most hard-nosed experimentalists are talking about integrating forty or more high-quality quantum bits (qubits) into a small programmable quantum computer—not in the remote future but in the next few years. If built, such a device will probably still be too small to do anything useful, but I honestly don't care.

The point is, forty qubits are enough to do *something* that computer scientists are pretty sure would take trillions of steps to simulate using today's computers. They'll suffice to disprove the skeptics, to show that nature really does put this immense computing power at our disposal—just as the physics textbooks have implied since the late 1920s. (And if quantum computing turns out *not* be possible, for some deep reason? To me that's unlikely, but even more exciting, since it would mean a revolution in physics.)

So, is imminent quantum supremacy the "most interesting recent [scientific] news"? I can't say that with any confidence. The trouble is, which news we find interesting depends on how widely we draw the circle around our own hobbyhorses. And some days, quantum computing seems to me to fade into irrelevance, next to the precarious state of the Earth. Perhaps when people look back a century from now, they'll say that the most important science news of 2015 was that the West Antarctic Ice Sheet was found to be closer to collapse than even the alarmists predicted. Or, just possibly, they'll say the most important news was that in 2015 the AI–risk movement finally went mainstream.

This movement posits that superhuman artificial intelligence is likely to be built within the next century, and that the biggest problem facing humanity today is to ensure that when the AI arrives, it will be "friendly" to human values (rather, than, say, razing the solar system for more computing power to serve its inscrutable ends). I like to tease my AI–risk friends that I'll be more worried about the impending AI singularity when my Wi-Fi stays working for more than a week. But who knows? At least this scenario, if it panned out, would render the melting glaciers pretty much irrelevant.

Instead of expanding my "circle of interest" to encompass the future of civilization, I could also contract it, around my fellow theoretical computer scientists. In that case, 2015 was the year that László Babai of the University of Chicago announced the first "provably fast" algorithm for one of the central problems in computing: graph isomorphism. This problem is to determine whether two networks of nodes and links are "isomorphic" (that is, whether they become the same if you relabel the nodes). For networks with n nodes, the best previous algorithm—which Babai also helped to discover, thirty years ago—took a number of steps that grew exponentially with the square root of n.

The new algorithm takes a number of steps that grows exponentially with a power of $\log(n)$ (a rate that's called "quasi-polynomial"). Babai's breakthrough probably has no applications, since the existing algorithms were already fast enough for any networks that would ever arise in practice. But for those who are motivated by an unquenchable thirst to know the ultimate limits of computation, this is arguably the biggest news so far of the 21st century.

Drawing the circle even more tightly, in "quantum query complexity"—a tiny subfield of quantum computing I cut my teeth on as a student—it was discovered this past year that there are Boolean functions that a quantum computer can evaluate in less than the square root of the number of input accesses that a classical computer needs, a gap that had stood as the record since 1996. Even if useful quantum computers are built, this result will have zero applications, since the functions that achieve this separation are artificial monstrosities, constructed only to prove the point. But it excited me: It told me that progress is possible, that the seemingly eternal puzzles that drew me into research as a teenager do occasionally get solved. So damned if I'm not going to tell you about it.

At a time when the glaciers are melting, how can I justify getting excited about a new type of computer that will be faster for certain specific problems—let alone about an artificial function for which the new type of computer gives you a slightly bigger advantage? The "obvious" answer is that basic research could give us new tools with which to tackle the woes of civilization, as it's done many times before. Indeed, we don't need to go as far as an AI singularity to imagine how.

By letting us simulate quantum physics and chemistry, quantum computers might spark a renaissance in materials science, and allow (for example) the design of higher-efficiency solar panels. For me, though, the point goes beyond that, and has to do with the dignity of the human race. If, in millions of years, aliens come across the ruins of our civilization and dig up our digital archives, I'd like them to know that before humans killed ourselves off, we at least managed to figure out that the graph-isomorphism problem is solvable in quasi-polynomial time and that there exist Boolean functions with superquadratic quantum speedups. So I'm glad to say that they *will* know these things, and that now you do, too.

A NEW ALGORITHM SHOWING WHAT COMPUTERS CAN AND CANNOT DO

JOHN NAUGHTON

Columnist, the *Observer*; Emeritus Professor of the Public
Understanding of Technology, Open University, U.K.; Emeritus
Fellow, Wolfson College, Cambridge;
author, *From Gutenberg to Zuckerberg*

The most interesting news came late in 2015—on November 10th, to be precise, when László Babai of the University of Chicago announced that he had come up with a new algorithm for solving the graph-isomorphism problem. This algorithm appears to be much more efficient than the previous "best" algorithm, which has ruled for over thirty years. Since graph isomorphism is one of the great unsolved problems in computer science, if Babai's claim stands up to the kind of intensive peer-review to which it is now being subjected, then the implications are fascinating—not least because we may need to rethink our assumptions about what computers can and cannot do.

The graph-isomorphism problem seems deceptively simple: how to tell when two different graphs (what mathematicians call networks) are really the same, in the sense that there's an "isomorphism"—a one-to-one correspondence between their nodes that preserves each node's connections—between them. Easy to state, but difficult to solve, since even small graphs can be made to look different just by moving their nodes around. The standard way to check for isomorphism is to consider all possible ways to match up the nodes in one network with those in the other. That's tedious but feasible for very small graphs, but it rapidly gets out of hand as the number of nodes increases. To compare two graphs with just ten nodes, for example, you'd have to check more than 3.6 million (i.e., 10 factorial) possible matchings. For graphs with 100 nodes, you're probably looking at a number bigger than all the molecules in the universe. And in a Facebook age, networks with millions of nodes are commonplace.

From the point of view of practical computing, factorials are really bad news, because the running time of a factorial algorithm can quickly escalate into billions of years. So the only practical algorithms are those for problems whose solutions can be expressed as polynomials (e.g., n-squared or n-cubed, where n is the number of nodes), because running times for them increase much more slowly than those for factorial or exponential functions.

The tantalizing thing about Babai's algorithm is that it is neither pure factorial nor polynomial but what he calls "quasi-polynomial." It's not clear yet what this means, but the fuss in the mathematics and computer-science community suggests that while the new algorithm might not be the Holy Grail, it is nevertheless significantly more efficient than what's gone before.

If that turns out to be the case, what are the implications? Well, first, there may be some small but discrete benefits. Babai's breakthrough could conceivably help with other kinds of computationally difficult problems. For example, genomics researchers have been trying for years to find an efficient algorithm for comparing the long strings of chemical letters within DNA molecules. This is a problem analogous to that of graph isomorphism, and any advance in that area may have benefits for genetic research.

But the most important implication of Babai's work may be inspirational—in reawakening mathematicians' interests in other kinds of hard problems that currently lie beyond the reach of even the most formidable computational resources. The classic example is the public-key encryption system on which the security of all online transactions depends. This works on *asymmetry*: It is relatively easy to take two huge prime numbers and multiply them together to produce an even larger number. But—provided the original primes are large enough—it is computationally difficult (in the sense that it would take an impracticable length of time) to factorize the product—that is, to determine the two original numbers from which it was calculated. If, however, an efficient factorizing algorithm were to be found, then our collective security would evaporate and we would need to go back to the drawing board.

DESIGNER HUMANS

MARK PAGEL

Professor of evolutionary biology, Reading University, U.K., author,
Wired for Culture

The use of CRISPR (Clustered Regularly Interspaced Short
Palindromic Repeats) technologies for targeted gene-editing
means that an organism's genome can be cheaply cut and then
edited at any location. The implications of such a technology
are potentially so great that "crisper" has become a widely heard
term outside science, being the darling of radio and television
talk shows. And why not? All of a sudden, scientists and bio-
technologists have a way of making designer organisms. The
technology's first real successes—in yeast, fish, flies, and even
some monkeys—have already been trumpeted.

But of course what is on everyone's mind is its use in humans.
By modifying genes in potential parental egg or sperm cells, it
will produce babies "designed" to have some desired trait (or
to lack an undesirable one). By editing genes early enough in
embryonic development—a time when only a few cells become
the progenitors of all the cells in our bodies—the same design
features can be obtained in the adult.

Just imagine: no more Huntington's chorea, no more sickle-
cell anemia, no more cystic fibrosis, or a raft of other heritable
disorders. But what about desirable traits—eye and hair color,
personality, temperament, even intelligence? The first of these
is already within CRISPR's grasp. The others are probably only
partly caused by genes and even then potentially by scores or

possibly hundreds of genes. But who's to say we won't figure out even those cases someday? The startling progress that genomic and biotechnological workers have made over the last twenty years is not slowing down, and there's reason to believe that (if not in our own lifetimes, surely in our children's) knowledge of how genes influence many of the traits we'd like to design into or out of humans will be widely available.

None of this is lost on the CRISPR community. Already there have been calls for a moratorium on the use of the technology in humans. But that was true in the early days of *in-vitro* fertilization, although not all were from the scientific community. The point is that our norms of acceptance of technological developments get shifted as those technologies become more familiar.

The current moratorium on the use of CRISPR technologies in humans probably won't last long. The technology is remarkably accurate and reliable and this is still early days. Refinements are inevitable, as are demonstrations of CRISPR's worth in ameliorating, say, agricultural or environmental problems. All this will wear down our resistance to designing humans. Already, CRISPR has been applied successfully to cultured cell lines derived from humans. The first truly and thoroughly designed humans are more than just the subjects of science fiction: They are on our doorstep, waiting to be let in.

CELLULAR ALCHEMY

ROGER HIGHFIELD

Director of external affairs, Science Museum Group, U.K.;
co-author (with Martin Nowak), *Supercooperators*

A remarkable convergence of technologies is sending shock-waves through genetics and medicine. The widespread adoption of easy-to-use, inexpensive, and effective genome-editing techniques has made recent headlines. But it's just as significant that their power has been amplified because we live in the era of cheap genome sequencing. There are ways to regulate how genes are used without making permanent changes to DNA, and we can now reprogram adult cells to return them to an embryonic state and then convert them into any desired cell type.

This is the era of cellular alchemy.

To underline why today's gene-editing methods are so important, look back to 1990, when John Clark's team at the Roslin Institute, near Edinburgh, unveiled Tracy the sheep. Tracy, the first transgenic farm mammal, was a pharming pioneer that made 30 grams of a human protein in every liter of her milk. Tracy was considered so significant that after her death in 1997, she was stuffed and placed in the Science Museum's collections.

She had been genetically altered to make alpha 1-antitrypsin (AAT), then regarded as a potential drug for the treatments of cystic fibrosis and emphysema. But the Roslin team could use only crude methods that offered no control over where the DNA would end up. They could add calcium salts to make DNA precipitate out of a solution onto the embryo, hoping that some would migrate inside; use electrical pulses to punch holes in the embryo membranes to drive DNA inside; package DNA in fat particles (liposomes), which dissolve in cell membranes; or, best of all, inject a few hundred copies of the DNA directly into the nucleus of a zygote. The Roslin team attached the AAT gene to the promoter region of a gene responsible for a sheep's-milk protein and injected 1,000 embryos with this "construct" to end up with one sheep—Tracy—that could make alpha-1-antitrypsin in her mammary gland.

The use of such crude genetic-engineering methods on people is inconceivable. But today we have a way to change DNA at a precise spot. Gene editing permits specific stretches of DNA to be deleted from genomes and new stretches to be inserted into the gap in a much more precise, reliable way. Gene-editing methods can also insert or remove a number of genes at a time, offering huge opportunities when it comes to altering crops, animals, and even people.

The reason this subtle knife is so powerful is that we wield it at a time when we can cheaply and easily sequence DNA to check edits. We know how to manipulate genes without altering them, by inducing epigenetic changes that regulate how they're used, and we know how to manipulate cells, too—notably by the use of "Yamanaka factors" to turn adult cells into embryonic cells.

When it comes to people, this will pave the way for model systems to test drugs, the creation of T-cells designed to fight cancer, creation of a patient's own disease-free cells for therapies, humanized pig organs, and so on. Work by Mitinori Saitou of Kyoto University and Azim Surani in the Gurdon Institute at Cambridge University has shown that these reprogrammed embryonic cells can even be turned into immortal germline cells.

By combining these technologies, one might take a skin biopsy from someone with a serious disease, correct the underlying genetic defect in these cells, convert them to primordial germ cells and thence into healthy, corrected sperm or eggs. Given the limitations of embryo screening, and assuming that the wider ethical concerns will be tempered by reasonable pragmatism, it is inevitable that one day children will be born with a skin cell as a parent. When that day dawns, the convergence of gene editing, sequencing, and reprogramming into cellular alchemy will have led to the permanent alteration of the human genome. With significant numbers of people born this way, human evolution will be heading on a new course.

A TERRIBLE BEAUTY HAS BEEN BORN

RANDOLPH NESSE
Foundation Professor of Life Sciences, director, Center for Evolution and Medicine, Arizona State University; co-author (with George C. Williams), *Why We Get Sick*

The biggest news of 2015 was recognition that new abilities to edit specific genes will transform life itself, transform it utterly. Simple enough to be implemented in labs everywhere, the CRISPR/Cas9 technique replaces a specified DNA sequence with a chosen alternative. The system has revolutionized genetic research and offers hope to those with genetic diseases. It also, however, enables us to change future generations. It even lends itself to creating "gene drives" that can, in just a few generations, replace a given sequence in all members of a sexually reproducing species. A terrible beauty has been born.

The possibilities are beyond our imagining, but some are already real. Trials with caged mosquitoes demonstrate fast transmission of a new gene providing resistance to malaria. If released into the wild, the gene could spread and eliminate that scourge. Would it? What else would it do? No one knows. Other gene drives could eliminate a species. Good riddance to smallpox, but what would happen to ecosystems without mice and mosquitoes? Again, no one knows.

Specters from Disney's *The Sorcerer's Apprentice* come to mind. How will a species with limited ability to control itself use such vast new power? The answer will determine the future of humankind and, indeed, life on this planet. In a remarkable demonstration of transparency and foresight, the National Academy of Sciences organized a meeting last December with sibling organizations from the U.K. and China to discuss the opportunities and threats. The risks were taken seriously, but there was little consensus on where this technology will take us and how (or if) we can control it. It will probably transform life itself, fast. In what way, no one knows.

DNA PROGRAMMING

PAUL DOLAN

Behavioral scientist, London School of Economics; author, *Happiness by Design*

At a time when groundbreaking discoveries seem almost commonplace, it is difficult to predict which scientific news is important enough to stay news for longer than a few days. To stick around, it would have to potentially redefine "who and what we are." One of the recent advances that fulfills these prerequisites is the decoding and reprogramming of DNA via bioinformatics.

Although mapping the human genome was a great achievement, it was bioinformatics that allowed a practical application of the acquired knowledge. Uploading a genome onto a computer has enabled researchers to use genetic markers and DNA amplification technologies in ways that shed light on the intricate, otherwise unfathomable gene/environment interactions causing disease. Researchers also hope to use bioinformatics to solve real-life problems; imagine microbes programmed to generate inexpensive energy, clean water, fertilizer, drugs, and food, or tackle global warming by sucking carbon dioxide from the air.

But like most things in life, there are also possible negative side effects. With DNA being written like software, cloning and designing complex living creatures, including humans, no longer seems a sci-fi fantasy. All possible advantages aside, it is likely to stir a wide range of ethical debates, requiring us to ponder what it means to be human: a naturally conceived, unique, and largely imperfect creature or a designed, aimed-to-be-perfect being?

Asked about the future of DNA coding and bioinformatics, genomics pioneer Craig Venter replied, "We're only limited by our imagination." I'm somewhat more skeptical, but one thing is for sure: Digital DNA is here to stay in scientific news about evolutionary biology, forensic science, and medicine, and also in debates about the way we humans define ourselves.

HUMAN CHIMERAS

DAVID HAIG

George Putnam Professor of Biology; Harvard University;
author, *Genomic Imprinting and Kinship*

A man conceived a child whose genotype did not match his own but was consistent with the child's being the grandson of the man's parents. The proposed explanation is that the man had a twin brother who was never born but whose cells colonized the man's testes when the man was a fetus. Those cells produced the sperm that conceived the child.

In the modern era of sensitive genetic testing, multiple examples of chimerism are being detected—where chimerism refers to a body containing cells derived from more than one fertilized egg. All of us probably contain replicating cells from more than one member of our genetic family. A distinction should be made between bodily individuals (who are chimeric) and genetic individuals who may be distributed across multiple bodies.

THE RACE BETWEEN GENETIC MELTDOWN AND GERMLINE ENGINEERING

JOHN TOOBY

Founder of evolutionary psychology; professor of anthropology,
UC Santa Barbara; codirector, UCSB Center for Evolutionary
Psychology

The most remarkable breaking news in science is that I exist.
Well, not just me. People like me who, without technology,
would have died early. Of the roughly 5.5 billion people who
have survived past puberty, perhaps only 1 billion would be here
were it not for modern sanitation, medicine, technology, and
market-driven abundance. Ancestrally, the overwhelming majority of humans died before they had a full complement of
children, often not making it past childhood. For those who
live in developed nations, our remodeled lifetables are among
the greatest of the humane triumphs of the Enlightenment—
delivering parents from the grief of holding most of their children dead in their arms, or of children losing their parents (and
then themselves dying from want).

But there is hidden and unwelcome news at the core of this
triumph. This arises out of the brutal way natural selection links
childbearing to the elimination of genetic disease.

The first thing to recall is that even our barest functioning depends on advanced organic technology at all scales—technology
engineered by selection. For example, our eyes—macroscopic
objects—have 2 million moving parts, and yet individual rods are

so finely crafted they can respond to single photons. Successful parents in every species live near summits on adaptive landscapes.

The second thing to remember is that physics is perpetually hurling us off these summits, assaulting the organization necessary to our existence. Entropy not only ages and kills us as individuals but also successfully attacks each parent's germline. Indeed, the real news is that a number of methods have converged on the estimate that every human child contains roughly 100 new mutations—genetic changes not present in their parents. To be sure, many of these occur in inert regions or are otherwise "silent" and so do no harm. But a few are harmful, and although the rest are individually small in effect, collectively they plague each individual with debilitating infirmities.

These recent estimates are striking when one considers how, in an entropy-filled world, we maintained our high levels of biological organization. Natural selection is the only physical process that pushes species' designs uphill—against entropy, toward greater order (positive selection)—or maintains our favorable genes against the downward pull exerted by mutation pressure (purifying selection). If a species is not to melt down under the hard rain of accumulating mutations, the rate at which harmful mutations are introduced must equal the rate at which selection removes them (mutation-selection balance). This removal is self-executing: Harmful genes cause impairments to the healthy design of the individuals they're situated in. These impairments (by definition) reduce the probability that the carrier will reproduce, and thereby they reduce the number of harmful genes passed on. For a balance to exist between mutation and selection, a critical number of offspring must die before reproduction— die because they carry an excess load of mutations.

Over the long run, successful parents average a little more

than two offspring that survive into parenthood. (The species would go extinct or fill the planet if the average were smaller or greater.) This is as true for humans with our handful of children as it is for an ocean sunfish with a nest full of 300 million. To understand how endlessly cruel the anti-entropic process of selection is, consider a sunfish mother with one nest. On average, 299,999,997 of her progeny die, and two or three become comparable parents. Since the genotypes of offspring are generated randomly, the number of coin flips (in 300 million series) guarantees a lower end of the bionomial distribution of mutations that are many standard deviations out. That is, there are two or three who become parents because they received a set of genes improbably free of negative mutations. These parents therefore restart the lineage's next generation, having shed enough of the mutational load to have rolled back mutational entropy to the parental level. Ancestral humans, with far smaller offspring sets, maintained our functional organization more precariously, having survived over evolutionary time on the edge of a far smaller selective gradient, between those children with somewhat smaller sets of impairments and those with somewhat larger sets. Most ancestral humans were fated by physics to be childless vessels whose deaths served to carry harmful mutations out of the species.

Now along comes the demographic transition—the recent shift to lower death rates and then lower birth rates. Malthusian catastrophe was averted, but the price of relaxing selection has been moving the mutation-selection balance toward an unsustainable increase in genetic diseases. Various naturalistic experiments suggest this meltdown can proceed rapidly. (Salmon raised in captivity for only a few generations were strongly outcompeted by wild salmon subject to selection.) Indeed, it is possible that the drop in death rates over the demographic transition caused, by increasing the genetic load, the subsequent drop in birth rates below replacement: If humans are equipped with physiological assessment systems to detect when they are in good enough condition to conceive and raise a child, and if each successive generation bears a greater number of micro-impairments that aggregate into, say, stressed exhaustion, then the paradoxical outcome of improving public health for several generations would be ever lower birth rates. One or two children are far too few to shed incoming mutations.

No one could regret the victory over infectious disease and starvation now spreading across the planet. But we as a species need an intensified research program into germline engineering, so that the Enlightenment science that allowed us to conquer infectious disease will allow us to conquer genetic disease (through genetic repair in the zygote, morula, or blastocyst). With genetic counseling, we have already focused on the small set of catastrophic genes, but we need to sharpen our focus on the extremely high number of subtle, minor impairments that statistically aggregate into major problems.

I am not talking about the ethical complexities of engineering new human genes. Imagine instead that at every locus, the infant received healthy genes from her parents. These would not be genetic experiments with unknown outcomes: Healthy genes are healthy precisely because they interacted well with each other over evolutionary time. Parents could choose to have children created from their healthiest genes, rather than leaving children to be shotgunned with a random and increasing fraction of damaged genes. Genetic repair would replace the ancient cruelty of natural selection, which fights entropy only by tormenting organisms because of their genes.

THE ONGOING BATTLES WITH PATHOGENS

ROBERT KURZBAN

Psychologist, University of Pennsylvania; founder and codirector,
Pennsylvania Laboratory for Experimental Evolutionary Psychology
(PLEEP); author, *Why Everyone (Else) Is a Hypocrite*

The end of the year saw two stories about pathogens, one hopeful and the other less so. The hopeful one is the advancing ability to insert genes that are desirable (from humans' point of view) into organisms and facilitate that gene's spread through the population.

Consider the case of malaria, a focus of current efforts. Malaria still infects tens of millions of people and causes hundreds of thousands of deaths. Inserting a gene that blocks malaria into a mosquito genome is helpful, but only to a limited degree: If the gene doesn't spread in wild populations, then its effects will be fleeting. If, however, mosquitoes are released that have the gene in question as well as genes that facilitate its spread, then the effects can be long-lasting.

The less hopeful case involves so-called "super-gonorrhea" strains of the pathogen, resistant to the antibiotics currently in use. As is well known, the use of antibiotics gives an advantage to resistant strains of pathogens. The tools we use to treat disease become the means by which we make the diseases harder to treat. Cases of super-gonorrhea have appeared in England, leading to a certain amount of concern.

Historically, as illustrated by the super-gonorrhea case, fighting pathogens has been something of an arms race: Humans build better weapons, pathogens evolve counter-strategies. But the present age is seeing a confluence of advances that promise to turn the arms race into something more like a winnable war. First is progress in genetics, including techniques for inserting genes into genomes. Second is a sophisticated understanding of evolution and a burgeoning of ideas about how to harness it. In the past, evolution worked against us—as in the case of resistant strains of pathogens; we are becoming better at harnessing it. And third is a growing sophistication in thinking about systems. It is unlikely that the same sorts of mistakes will be made as were made in the past, when simplistic ecological interventions led to disastrous outcomes, as in the case of cane toads in Australia. Our view of ecosystems is both humbler and more sophisticated than ever before.

It seems likely that pathogens of one sort or another will be important for a long time: They are moving targets with tremendous capacity for harm. On the other hand, recent advances may well be able to tame them in ways we have not previously seen.

ANTIBIOTICS ARE DEAD; LONG LIVE ANTIBIOTICS!

AUBREY DE GREY

Gerontologist; chief science officer, SENS Foundation;
author, *Ending Aging*

We've been hearing the tales of doom for quite a few years now:
The breathtaking promiscuity of bacteria, which allows them to
mix and match their DNA with that of others to an extent that
puts Genghis Khan to shame, has allowed them to accumulate
genetic resistance to more and more of our antibiotics. It's been
trumpeted for decades that the rate at which this occurs can
be slowed by careful use, especially by not ceasing a course of
antibiotics early—but inevitably there is lack of compliance, and
here we are with MRSA, rife in hospitals worldwide, and other
major species becoming more broadly antibiotic-resistant with
every passing year. The bulk of high-profile expert commentary
on this topic is becoming more dire with every passing year.

But this pessimism rests on one assumption: that we have no
realistic prospect of developing new classes of antibiotics anytime soon—antibiotics that our major threats have not yet seen
and thus not acquired resistance to. And it now seems that that
assumption is unwarranted. It is based the fact that no new antibiotic class with broad efficacy has been identified for decades.
But recently a novel method was identified for isolating exactly
those, and it seems to work really, really well.

It arose from a case of sheer chutzpah. Scientists from Boston
and Germany got together and reasoned as follows:

1. Antibiotics are generally synthesized in nature by bacteria (or other microbes) as defenses against each other.
2. We have identified antibiotics in the lab, and thus necessarily only those made by bacterial species that we can grow in the lab.
3. Almost all bacterial species cannot be grown in the lab using practical methods.
4. That hasn't changed for decades.
5. But those bacteria grow fine in the environment, typically the soil.
6. So can we isolate antibiotics from the soil?

And that's exactly what they did. They built a device that let them isolate and grow bacteria in the soil itself, with molecules freely moving into and out of the device, thereby sidestepping our ignorance of which such molecules actually matter. They were able to isolate the compounds those bacteria were secreting and test them for antibiotic potency. And it worked. They found a completely new antibiotic, which has been shown to have broad efficacy against several bacterial strains resistant to most existing antibiotics.

As if that were not enough, here's the kicker. This was not some kind of massive high-throughput screen of the kind we so often hear about in biomedical research these days. The researchers tried this approach just once, in essentially their backyard, on a very small scale, and it *still* worked the first time. What that tells us is that it can work again—and again, and again.

Don't get me wrong. There is certainly no case for complacency at this stage. This new compound and those discovered by similar means will still need to grind their way through the usual process of clinical evaluation—though there is reason for considerable optimism that that process is speeding up, with the recent case of an Ebola vaccine being a case in point. But still, even though any optimism must for now be cautious, it is justified. Pandemics may not be our future after all.

THE 6 BILLION LETTERS OF OUR GENOME

ERIC TOPOL, M.D.

Professor of genomics, Scripps Research Institute; director, Scripps
Translational Science Institute; author, *The Patient Will See You Now*

In 2015, we crossed a threshold: The first million people had
their genomes sequenced. Beyond that, based on progress in
sequencing technology, it is projected that we'll hit 1 billion
people sequenced by 2025. That seems formidable but likely,
given that the pace of DNA-reading innovation has exceeded
Moore's Law. The big problem is not amassing billions of
human genome sequences but how to understand the significance of each of the 6 billion letters that comprise our genome.

About 98.5 percent of our genome is not made of genes
and so doesn't directly code for proteins. But most of this non-
coding portion of the genome influences, in one way or another,
how genes function. While it's relatively straightforward to understand genes, the non-coding elements are far more elusive.

So the biggest breakthrough in genomics—*Science's* 2015
Breakthrough of the Year—is the ability to edit a genome, via
so-called CRISPR technology, with remarkable precision and
efficiency. We have had genome-editing technologies for several years, including zinc-finger nucleases and TALENs [tran-
scription activator-like effector nucleases], but they weren't easy
to use, nor could they achieve a high rate of successful editing
in the cells that were exposed. The precision problem also extended to the need to avoid editing in unintended portions of

the genome (so-called off-target effects). Enter CRISPR, and everything has quickly changed.

Many genome-editing clinical trials are now under way, or will be soon, to treat medical conditions for which treatment or a cure has proved remarkably challenging. These include sickle-cell disease, thalassemia, hemophilia, HIV/AIDS, and some very rare metabolic diseases. Indeed, the first person whose life was saved (by TALENs) was a young girl with leukemia who had failed all therapies attempted until she had her T-cells genome-edited. George Church and his colleagues at Harvard were able to edit sixty-two genes of the pig's genome to make it immunologically inert, so that the idea of transplanting an animal's organ into humans—xenotransplantation—has been resurrected. A number of biotech and pharma companies (Vertex, Bayer, Celgene, and Novartis), have recently partnered with the editing-company startups (CRISPR Therapeutics, Editas Medicine, Intellia Therapeutics, Caribou Biosciences) to rev up clinical programs.

But the biggest contribution of genome editing, and specifically that of CRISPR, is to catapult the field of functional genomics forward. Not understanding the biology of the DNA letters is the biggest limitation of our knowledge base in the field. Many interesting DNA-sequence variant "hits" have been discovered but overshadowed by uncertainty. Determining functional effects of the VUS (variants of unknown significance) has moved at a sluggish pace, with too much of our understanding of genomics based on population studies rather than on pinpointing the biology and potential change in function due to an altered (compared with the reference genome) DNA letter.

Now we've recently seen how we can systematically delete genes to find out which are essential for life. From that, we learned that only about 1,600 (8 percent) of the nearly 19,000 human genes are truly essential. All of the known genes implicated in causing or contributing to cancer can be edited—and, indeed, that systematic assessment is well under way. We have just learned how important the 3D structure of DNA is for cancer vulnerability, by using CRISPR to edit out a particular genomic domain. Moreover, we can now generate a person's cells of interest (from their blood cells, via induced pluripotent stem cells)—to make heart, liver, brain, or whatever the organ/tissue of interest. When this is combined with CRISPR editing, it becomes a remarkably powerful tool that takes functional genomics to an unprecedented level.

What once was considered the "dark matter" of the genome is about to be illuminated. The greatest contribution of genome editing will ultimately be to understand the 6 billion letters that comprise our genome.

SYSTEMS MEDICINE

STUART A. KAUFFMAN

Theoretical biologist; founding director, Institute for Biocomplexity and Informatics, University of Calgary, Canada; author, *Humanity in a Creative Universe*

Systems Medicine is emerging, a new holistic view of the organism and the integrated molecules, cells, tissues, and organs comprising that organism living in its world. We are heritors of over forty years of wonderful molecular biology, which was, however, somewhat overconfident of a molecular reductionism that failed to integrate the pieces.

Within each cell is a vast genetic regulatory system coordinating the activities of thousands of genes—that is, which genes are transcribed when and where, along with new knowledge about epigenetic factors such as histone modifications. These comprise a complex nonlinear dynamical system whose coordinated behaviors, coupled with the physics and chemistry of molecules and structures within and between cells and the environment, mediate ontogeny and disease.

It is now becoming known that some of these genetic factors form autoregulatory feedback loops, which are likely to underlie alternative dynamical "attractors," or stable alternative patterns of gene expression, underlying different cell types. The idea of cell types as alternative attractors goes back to Nobel laureates Jacob and Monod in 1963. If cell types are such attractors, each drains a "basin of attraction" in its state space. Then cell differentiation is a flow among attractors induced by signals or noise, or "bifurcations" to new attractors, as parameters change. Not only cells but tissues and organs may be nonlinear dynamical systems with attractors linked hierarchically in unknown ways.

This fine, if early, holistic dynamical picture leaves out the myriad biological functions of these variables. We need an enhanced physiology of the total organism in its world. We live in environments. Odd chemicals can switch an antenna to a leg in genetically normal developing fruit flies. What of the thousands of new chemicals unleashed into the atmosphere?

How can we control and try to "treat" such complex systems? Think of a spring mattress, with linked springs all wiggling. Now, would you try to control the wiggling springs by throwing a small pillow on one spring? Not often, unless its unique product directly mediated a disease. You would try to subtly alter the wiggling of the springs to get the coordinated behavior you want. The same applies to us as patients with vastly complex nonlinear systems underlying health and disease. We need to begin carefully to move toward combinatorial therapies (or multiple pillows), a move that is gradually happening. This move may require new testing procedures beyond our current gold standard of randomized clinical trials, which really only work well if the many factors involved each affect the "phenotype" independently. This is rare in biology, where causality is multiple and interwoven, with feedback loops in complex networks with complex topology and "logic."

But there is hope: We can empirically climb "clinical fitness landscapes," each described with many variables, where peaks represent good treatments by one or many variables, from one or a set of drugs to environmental factors. In fact, almost anecdotal evidence, a kind of "learning by doing," can search such rugged clinical landscapes. More, Bayesian and other models of the underlying multi-causal mechanisms can guide our empirical search.

It is a time of hope as we step toward a holistic view of the organism in its world.

GROWING A BRAIN IN A DISH

SIMON BARON-COHEN

Professor of developmental psychopathology, director, Autism
Research Centre, Cambridge University; author, *The Science of
Evil* and *The Essential Difference*

One morning three years ago, my talented PhD student
Dwaipayan Adhya (known affectionately in the lab as Deep)
came into my office. He looked me straight in the eye and said
he'd like to grow autistic and typical neurons in a dish, from the
earliest moment of development, to observe how the autistic
neuron differs from the typical neuron day by day. I dropped
everything and listened.

Sounds like science fiction? You might imagine that to grow
a brain cell in a dish the scientist would first have to pluck a
neuron from a human embryo, keep it alive in a petri dish, and
then watch it under the microscope, measuring how it grows
day by day. If that's what you're imagining, you're wrong. There
is no way to get a neuron from a human embryo in any ethically
acceptable way, for obvious reasons.

So what method was Deep planning to use? He told me
about Shinya Yamanaka of Kyoto University, awarded the Nobel
Prize in 2012 (with Cambridge scientist John Gurdon) for his
work on induced pluripotent stem cells, or iPSC. In the lab, we
call this magic. Here's how it works.

Pluck a hair from the head of an adult, then take the follicle from that hair and, using the Yamanaka method, *reverse* the
cell, backward from the adult hair follicle into the state of a

stem cell—that is, back to being an undifferentiated cell before it became a hair follicle. This is not an embryonic stem cell; it is an "induced pluripotent" stem cell. Induced because the scientist has forced the adult hair follicle (though you could use any cell in the body), by genetic reprogramming, back into the stem-cell state. And "pluripotent" means it can now be genetically programmed to become any kind of cell in the body—an eye cell, a heart cell, a neuron. If the last one, this is referred to as "neuralizing" the iPSC.

I said to Deep, "Let's do it!" It seems entirely ethical: Most adults would be happy enough to donate a single hair from their head; no animal is "sacrificed" in this kind of science; and it enables scientists to study the development of human neurons in the lab.

The importance of Yamanaka's scientific breakthrough is that if you want to study development from the first moment of life, iPSC bypasses the need for an embryo. Before this, if you wanted to understand the autistic brain, scientists would rely on postmortem studies when the next of kin donated their autistic relative's brain to scientific research.

Brain donations are invaluable, but from a scientific perspective, postmortem brain tissue has many limitations. For example, you may end up with a set of brains donated from individuals of different ages, each of whom died from different causes. Interpretation of results thus becomes difficult. A further complication is that you may know very little about the deceased (e.g., what their IQ was or what their personality was like) and it is often too late to gather such information. Postmortem studies are still informative but come with a handful of caveats.

Alternatively, if you want to study the autistic brain, you can use an animal model; for example, you create a "knockout" mouse—a mouse genetically engineered to lack a particular gene that you suspect may play a role in autism—and observe its behavior compared to a typical (or wild-type) mouse. If the knockout mouse shows "autistic" behavior—for example, being less sociable—you conclude that the gene that was removed may be causing one or other of the symptoms of human autism. You can see the limitations of such animal studies immediately: How do you know that sociability in a mouse is the same thing as sociability in a human? The interpretation of results from such animal experiments is as littered with caveats as the postmortem studies.

Now we can see the power of adding iPSC to the scientist's toolkit for getting answers to questions. If you want to observe the living human brain, you can study the brain from the person you're interested in and gather as much information about that person as you want: IQ, personality, precise diagnosis—anything else you want. You can even look at the effects of different drugs or molecules on the neuron without having to do these arguably unethical drug studies on an animal.

The technique is not without its own limitations. An iPSC may not be exactly identical to an embryonic stem cell, so the neuralized iPSC may not be exactly the same as a naturally growing neuron. All tools in the scientist's toolkit have their limitations, but this one—to my mind—is more ethical and more directly relevant to autism than is animal research. Many labs, like ours, are testing whether you get the same results from both iPSC and postmortem studies, since this strengthens the conclusions that can be drawn.

Deep's exciting results will be published in 2016. The combination of a breakthrough scientific method in the hands of a talented young PhD student might just be a game-changer in our understanding of the causes of autism.

SELF-DRIVING GENES ARE COMING

STEWART BRAND

Founder, *The Whole Earth Catalog*; cofounder, *The Well*; cofounder,
The Long Now Foundation; author, *Whole Earth Discipline: An
Ecopragmatist Manifesto*

The new biotech tool called "gene drive" changes our relation
to wild species profoundly. Any gene (or set of genes) can be
forced to "drive" through an entire wild population. It doesn't
matter if the trait the genes control is deleterious to the organism. With one genetic tweak to its germline, a species can even
be compelled to go extinct.

The technique works by forcing homozygosity. Once the
genes for a trait are homozygous (present on both chomosomes)
and the parents are both homozygous, they will breed true for
that trait in all their descendants. Artificially selecting for desired traits via homozygosity is what breeders do. Now there's a
shortcut.

In effect, gene-drive genes forbid the usual heterozygosity in
cross-bred parents. In any two parents, if one of them is gene-
drive homozygous, all their offspring will be gene-drive homozygous and will express the gene-drive trait. Proviso: it only
works with sexually reproducing species—forget bacteria. And
it spreads quickly enough only in rapidly reproducing species—
forget humans.

The mechanism was first described as a potential tool by
Austin Burt of Imperial College London, in 2003. The way it
works is that a "homing endonuclease gene" cuts the DNA in

the adjoining chromosome and provides the template for the DNA repair, thus duplicating itself. In Richard Dawkins's terms, it is an exceptionally selfish gene. Heterozygous becomes homozygous, and after several generations the gene is present in every individual of the population. The phenomenon is common in nature.

Gene drive shifted from an interesting concept to a powerful tool with the arrival in the last few years of a breakthrough in genome editing called CRISPR/Cas9. Suddenly genes could be edited easily, cheaply, quickly, and with great precision. It was a revolution in biotech.

In 2014, George Church and Kevin Esvelt at Harvard published three papers spelling out the potential power of CRISPR-enabled gene drive and the kind of public and regulatory oversight needed to ensure its responsible deployment. They also encouraged the development of an "undo" capability. Ideally the effects of an initial gene-drive release could, if desired, be reversed before it spread too far, with the release of a countermanding secondary gene drive.

The benefits of gene drive could be huge. Vector-borne scourges like malaria and dengue fever could be eliminated by eliminating (or just adjusting) the mosquitoes that carry them. Food crops could be protected by reversing herbicide-resistance in weeds. Wildlife conservation would be free of one of its worst threats—the alien invasive rats, mice, ants, etc. that destroy native species on ocean islands. With gene drive, the invaders could be extirpated (driven extinct locally), and the natives would be protected permanently.

Developments are coming quickly. A team at Harvard proved that gene drive works in yeast. A team at UC San Diego inadvertently proved that it works in fruit flies. Most important,

Anthony James at UC Irvine and colleagues showed that malaria mosquitoes could be altered with gene drive so that they no longer carry the disease. Kevin Esvelt is developing a project to do the same with white-footed mice, the wildlife reservoir for Lyme disease in humans; if they're cured, humans will be as well.

The power to permanently change wild populations genetically is a serious matter. There are ecological questions, ethical issues, and many technical nuances that have to be examined thoroughly. Carefully, gradually, they will be.

Humanity has decided about this sort of thing before. Guinea worms are a horrible parasite that once afflicted 2.5 million people, mostly in Africa. In 1980, disease-control experts set about eliminating the worms from the world, primarily through improved water sanitation. That goal of deliberate extinction is now on the brink of completion. One of the strongest advocates of the project, President Jimmy Carter, declared publicly, "I would like the last Guinea worm to die before I do."

Gene drive is not a new kind of power, but it is a new level of power. And a new level of responsibility.

LIFE DIVERGING

JUAN ENRIQUEZ

Managing director, Excel Venture Management; co-author (with
Steve Gullans), *Evolving Ourselves*

The two rules governing what lives and what dies in the long
term have been pretty clear: natural selection and random mutation. But over the last century or two, and especially over the
last decade, humans have fundamentally altered those rules. Life
as we know it will undergo rapid and accelerating change; it will
will diverge, especially post–May 2014.

Already we largely determine what lives and dies on half
the surface of the planet—anywhere we've built cities, suburbs,
parks, farms, ranches. That makes cornfields and gardens alike
some of the world's most unnatural places. Nothing lives and
dies there except what we want, where and when we want. Orderly rows of plants that please us. All else is culled. (But leave it
fallow and untended for a couple of years and you will begin to
see what is driven by natural selection.)

In redesigning our environment, we create and nurture unnatural creatures: miniature pigs the size of Chihuahuas; corn that
cannot self-replicate; the big tom turkeys to dine on at Thanks-
giving, animals so grossly exaggerated that they can't copulate and
require artificial insemination. Today's big-breasted beasts are on
average 225 percent larger than they were in the 1930s.

Without human intervention, most of the creatures that live
around you would have been selected out. (Let a Lhasa Apso
loose on the African plain and watch what happens.) Same is

true of humans: In an all-natural environment, most of humanity would not be alive. But unnaturally selecting out microbes and viruses—like smallpox, polio, bubonic plague, and most infections—means billions of us get to live.

As we practice extreme human intervention and alter the course of natural selection, we create a parallel evolutionary track, one whose rules and outcomes depend on what we want. Life begins to diverge from what nature would design and reward absent our conscious and unconscious choices. A once unusual observation—that during the Industrial Revolution black moths in London survived better than white moths because they were better camouflaged in a polluted environment—has become the norm. Life around us is now primarily black-moth-like adaptations: cute dogs, cats, flowers, foods. We have altered plants, animals, and bacteria so extensively that to survive they have to reward us, or at least be ignored by us.

These two parallel evolutionary systems—one driven by nature, the other by humans—expand diverging evolutionary trees. The divergence between what nature would choose and what we choose gets ever larger. Many of the life-forms we are so accustomed to and dependent on would disappear or radically modify in our absence. But the true breakpoint began over the past few decades, when we began not just choosing how to breed but rewriting the code of life itself. In the 1970s and 1980s, biotechnology allowed us to insert all kinds of gene instructions. Random mutation is gradually being displaced by intelligent design. By 2000, we were decoding entire genomes and applying this knowledge to alter all kinds of life-forms. Today's high schoolers can spend $500 and alter life code using methods like CRISPR; these types of technologies can alter all subsequent generations, including humans.

In May 2014, a team of molecular biologists led by Floyd Romesberg created a new genetic code, a self-replicating system that codes life-forms using chemically modified DNA. Insofar as we know, for almost 4 billion years all life on this planet replicated using the four known base pairs of DNA (A, T, C, and G). Now we can swap in other chemicals. This third evolutionary logic-tree of life would initially be completely human-design-driven and could rapidly diverge from all known life. In theory, scientists could begin breeding plants and animals with a very different genetic makeup from that of any other creature on the planet. And these new life-forms may be immune to all known viruses and bacteria.

Moreover, if we discover life on other planets, something that seems likely, the biochemistry of these other life-forms will doubtless further increase the variety of life on Earth, providing life designers with new instruments and ideas to program/redesign existing life-forms so they can adapt to different environments.

Thus the biggest story of the next few centuries will be how we begin to redesign life-forms, spread new ones, develop approaches and knowledge to further push the boundaries of what lives where. And as we deploy all this technology, we will see an explosion of new forms that could make the Cambrian explosion look tame.

Life is expanding and diverging. Humans won't be immune to this trend. We've already coexisted and interbred with other versions of hominins; it was normal and natural for different versions of ourselves to be walking around. Soon we might return to this historically normal state, but with far more, and perhaps radically different, versions of ourselves. All of which may lead to just a few ethical, moral, and governance challenges.

FUNDAMENTALLY NEWSWORTHY

STUART FIRESTEIN

Chairman, Department of Biological Sciences, Columbia University;
author, *Failure: Why Science Is So Successful*

The all-consuming news story in biology last year—this decade, really—is the discovery of the CRISPR/Cas9 system and its practical application for gene editing. There have been numerous articles in the popular press. Most of that attention has been directed at the tremendous and potentially dangerous power of this new technology: It allows editing the DNA of genomes, including those of humans, in a way that would be permanent—that is, heritable through generations.

All this attention on the possible uses and misuses of CRISPR/Cas9 has obscured the real news—which is, in a way, old news. CRISPR/Cas9 is the fruit of years of fundamental research conducted by a few dedicated researchers who were interested in the arcane field of bacterial immunity. Not immunity to bacteria, as you might at first think, but how bacteria protect themselves against attack by viruses. Weird as it may seem, there are viruses that specialize in attacking bacteria. Just as most viruses we know about are specific to one species or another (you can't catch a cold from your dog), there are viruses that infect only bacteria—in fact, only certain types of bacteria. These have a special name: phage. And they have a long history in the development of molecular biology and genetics. Indeed, molecular biology began with the study of phage and its ability to insert its genome into the genomes of bacteria—even before Watson and Crick's famous articulation of DNA as the molecule of heredity.

For the past forty years, restriction enzymes, another family of bacterial proteins, have been the mainstay of the biotech industry, and they too were discovered first as an early example of bacterial protection mechanisms. And they were also discovered in university research laboratories devoted to basic research. CRISPR/Cas9, however, is a more sophisticated mechanism, approaching that of the immune system of higher animals: It is adaptive, in the sense that a bacterium and the bacteria it generates by dividing can "learn" and destroy the DNA of the genome of a particular type of phage after the bacterium has been attacked by that phage once. The researchers who discovered CRISPR/Cas9 and recognized its potential value as a gene-editing tool in living things other than bacteria were not searching for some new technology; they were after a deeper understanding of a fundamental question in prokaryotic (microbial) biology and evolution—the back-and-forth competition between bacteria and the viruses that invade them. Could that be any more arcane-sounding?

It is also important to recognize that this was not a serendipitous discovery, a happy accident along the way. This is often the case made for supporting fundamental research: You never know where it might lead; serendipity intervenes so often. But this is a false conception, and CRISPR/Cas9 is a perfect example of why. This was no simple accident resulting from good luck or happenstance. It was the fruit of hard and sustained labor, of whole careers devoted to understanding the fundamental principles of life. The particular groups that discovered CRISPR/Cas9 were looking for precisely such an adaptive, immune-like response in bacteria. Understanding the value of restriction enzymes, as these researchers would have, was a sensible step to appreciating the value of an even more sophisticated DNA-based protective system. This is how research works—neither by accident nor purposefully: It is the result of hard work at every level of inquiry. Advances are indeed often unpredictable, but that doesn't make them merely lucky. Certainly not like winning some type of lottery.

We continue to have this misguided debate about fundamental versus applied research as if they were two spigots that could be operated independently. They are one pipeline, and our job is to keep it flowing. This is old news, but we should never tire of stating it.

PALEO-DNA AND DE-EXTINCTION

W. TECUMSEH FITCH
Professor of cognitive biology, University of Vienna;
author, *The Evolution of Language*

When prehistoric humans arrived in America, they found a continent populated by mammoths, woolly rhinos, giant sloths, saber-toothed cats, horses, and camels. By the time Columbus arrived, all those species were extinct, due mainly to human hunting. But today paleobiologists are sequencing their genomes. Furthermore, genetic engineering is approaching the point where genetically engineered versions of those extinct species may walk the Earth again. Last year's big news—cloning of a dead pet dog—is a small beginning.

In the near future, the news will concern what paleo-DNA specialist Beth Shapiro dubs "de-extinction": generating living organisms bearing genes recovered from extinct species. Paleo-DNA (the somewhat degraded DNA recovered from bones or hair of extinct species) can be extracted from extinct species like mammoths and those exterminated by humans during historical times, including passenger pigeons, dodos, and thylacines (Tasmanian wolves). Trace amounts of DNA can be recovered, amplified, and sequenced. Key genes could then be engineered into cells of the closest living relatives (Asian elephants, for mammoths) to produce shaggy, cold-tolerant elephants. This is nearly within our technological reach. (*Jurassic Park* fans, take note: The truly ancient DNA from dinosaurs is too degraded to currently allow sequencing.) Although birds pose unique challenges—

200

being unclonable with current technologies—the de-extinction of passenger pigeons and moas (3.5-meter-tall flightless birds exterminated by hunting when the Maori arrived in New Zealand) appears within our grasp, and major de-extinction projects for these species are already under way.

So should we do this? Shapiro's recent book, *How to Clone a Mammoth,* provides an excellent introduction to the arguments. Given the polarized opinions generated by reintroducing wolves to Yellowstone, one can easily imagine the diversity of public reactions to reintroducing saber-toothed tigers or giant cave bears. The best pro arguments are ecological: By reviving lost species, we can restore habitats damaged or destroyed as our own species spread around the planet. Mammoth-like elephants stomping through the tundra of Siberia's Paleozoic Park would benefit the environment by slowing the process of permafrost melting and the attendant carbon release. From a purely scientific, curiosity-driven viewpoint, de-extinction will offer biological and ecological insights available in no other way. Tourists would pay top dollar to watch moas wandering the beech forests of New Zealand, or mammoths roaming Siberia. The con arguments are mostly practical (Why spend money reviving extinct species rather than saving living endangered species?) or techno-fearful (humans shouldn't play God), but no less passionately advocated.

By far the most significant issues will concern extinct hominids, like Neanderthals, and society needs to prepare for the challenging ethical questions raised by such research. In 1997, researchers at Svante Pääbo's paleo-DNA lab in Leipzig made the news by sequencing mitochondrial DNA from Neanderthals. Today, after breathtaking technological progress, a full Neanderthal genome is available online. Even more exciting, in 2010 Pääbo's group discovered Denisovans, a previously unknown Asian hominid species, based on DNA extracted from a finger bone. The discovery of Denisovans from paleo-DNA makes it crystal clear that when modern humans emerged from Africa, they encountered a world inhabited by multiple near-human species—all of them now extinct.

Recovering the genome sequence of an extinct hominid species is exciting because it provides answers to a host of biological questions that stones and bones—the previous mainstay of paleoanthropology—will never answer. For example, it seems likely, based on pigmentation genes, that Neanderthals had light skin and some had red hair. Paleo-DNA has clarified that occasional interbreeding probably occurred when the first modern humans migrated out of Africa and encountered Neanderthals. As a result, all non-African human populations bear traces of Neanderthal DNA in their genomes (and many Asians bear additional Denisovan DNA). Similarly, the issue of whether Neanderthals had spoken language has divided scholars for decades. Although the case remains far from closed, we now know that Neanderthals shared the derived human version of the FOXP2 gene, which enhances our speech motor control. This suggests that Neanderthals could at least produce complex vocalizations, even if they lacked modern, syntactic language. Such findings have fueled an ongoing sea change in contemporary reinterpretations of Neanderthals—from oafish thugs to smart, resourceful near-humans.

Paleo-DNA sequencing has changed our understanding not just of Neanderthals but of ourselves. Neanderthals were not modern humans: They lacked the rapid cultural progress characterizing our species, and thus presumably some of our cognitive capacities. But what precisely were these differences? Do these differences make us, the survivors, "human?" Or were

Neanderthals human but "differently abled?" Certainly, with the bodies of Olympic wrestlers and brains slightly larger than those of modern humans, they'd be first picks for your rugby scrum; perhaps they had unique cognitive abilities as well. Progress will be rapid in addressing these issues, because each new insight into the genetic basis of the human brain yields parallel insights into our understanding of Neanderthal brains—and of the cognitive differences between the two.

But the deep ethical issues concern the possible de-extinction of Neanderthals or other extinct hominids. From a scientific viewpoint, this would promise insights into hominid evolution and human nature unimaginable a decade ago. But from a legal viewpoint it would involve creating humans expressing Neanderthal genes, and thus require human cloning, already forbidden in many countries. But few doubt that within this century genetic engineering of our own species will be both technologically possible and ethically acceptable in at least some subcultures. Clearly, a human expressing Neanderthal genes (as many of us already do!) would retain all basic human rights, but the moral and ethical implications raised by Neanderthals in the workplace (or on college football teams) might easily eclipse those raised by racism or slavery.

Clearly, paleo-DNA will remain in the news for the foreseeable future, offering scientific insights and posing unprecedented ethical quandaries. It thus behooves all thinking people (especially politicians drafting legislation) to get acquainted with the technology and the biological facts before forming an opinion.

THE WISDOM RACE IS HEATING UP

MAX TEGMARK

Theoretical physicist and cosmologist, MIT; scientific director, Foundational Questions Institute; cofounder, Future of Life Institute; author, *Our Mathematical Universe*

There's a race going on that will determine the fate of humanity. Just as it's easy to miss the forest for all the trees, however, it's easy to miss this race for all the scientific news stories about breakthroughs and concerns. What do these headlines from 2015 have in common?

AI Masters 49 Atari Games Without Instructions

Self-driving Car Saves Life in Seattle

Pentagon Seeks $12-$15 Billion for AI Weapons Research

Chinese Team Reports Gene-Editing Human Embryos

Russia Building Dr. Strangelove's Cobalt Bomb

They are all manifestations of the race heating up—the race between the growing power of technology and the wisdom with which we manage it. The power is growing because our human minds have an amazing ability to understand the world and to convert this understanding into game-changing technology. Technological progress is accelerating for the simple reason that breakthroughs enable other breakthroughs: As technology gets twice as powerful, it can often be used to design and build technology that is twice as powerful in turn, repeated capability doubling in the spirit of Moore's Law.

What about the wisdom ensuring that our technology is beneficial? We have technology to thank for all the ways in which today is better than the Stone Age, but this is not only thanks to the technology but also to the wisdom with which we use it. Our traditional strategy for developing such wisdom has been learning from mistakes: We invented fire, then realized the wisdom of having fire alarms and fire extinguishers. We invented the automobile, then realized the wisdom of having driving schools, seat belts, and airbags.

For a while, it was OK for wisdom to lag behind in the race, because it would catch up when needed. With more powerful technologies such as nuclear weapons, synthetic biology, and future strong artificial intelligence, however, learning from mistakes is not a desirable strategy. We want to develop our wisdom in advance, so that we can get things right the first time, because that might be the only time we'll have. In other words, we need to change our approach to tech risk from reactive to proactive. Wisdom needs to progress faster.

The latest *Edge* Question is cleverly ambiguous and can be interpreted either as a call to pick a news item or as a query about what constitutes interesting and important news. If we define "interesting" in terms of clicks and Nielsen ratings, then top candidates must involve sudden change of some sort, whether it be a discovery or a disaster. If we instead define "interesting" in terms of importance for the future of humanity, then our top list should include developments too gradual to meet a journalist's definition of "news"—such as "Globe Keeps Warming." In that case, I'll put the heating up of the wisdom race at the top of my list. Why?

From my perspective as a cosmologist, something remarkable has just happened: After 13.8 billion years, our universe has finally awoken, with small parts of it becoming self-aware, marveling at the beauty around them and beginning to decipher how their universe works. We, these self-aware life-forms, are using our newfound knowledge to build technology and modify our universe on ever grander scales.

This is one of those stories where we get to pick our own ending, and there are two obvious ones for humanity to choose between: Either win the wisdom race and enable life to flourish for billions of years, or lose the race and go extinct. To me, the most important scientific news is that after 13.8 billion years we finally get to decide—probably within centuries or even decades.

Since the decision about whether to win the race sounds like a no-brainer, why are we still struggling with it? Why is our wisdom for managing technology so limited that we didn't do more about climate change earlier? Why have we come close to accidental nuclear war more than a dozen times? As Skype founder Jaan Tallinn likes to point out, it's because our incentives drove us to a bad Nash equilibrium. Many of humanity's most stubborn problems, from destructive infighting to deforestation, overfishing, and global warming, have this same root cause: When everybody follows the incentives they're given, it results in a worse situation than cooperation would have enabled.

Understanding this problem is the first step toward solving it. The wisdom we need to avoid lousy Nash equilibria must be developed at least in part by the social sciences, to help create a society wherein individual incentives are aligned with the welfare of humanity, encouraging collaboration for the greater good. Evolution endowed us with compassion and other traits to foster collaboration, and when increasingly complex technology made these evolved traits inadequate, our forebears developed peer pressure, laws, and economic systems to steer their societies toward good Nash equilibria. As technology gets ever more powerful, we need ever stronger incentives for those who develop, control, and use it to make its beneficial use their top priority.

Although the social sciences can help, plenty of technical work is needed in order to win the race. Biologists are now studying how to best deploy (or not) tools such as CRISPR genome editing. 2015 will be remembered as the year when the beneficial AI movement went mainstream, engendering productive symposia and discussions at all the largest AI conferences. Supported by many millions of dollars in philanthropic funding, large numbers of AI researchers around the world have begun investigating the fascinating technical challenges involved in keeping future AI systems beneficial. Thus has the laggard in the all-important wisdom race gained significant momentum in 2015! Let's do all we can to make future news stories be about wisdom winning the race, because then we all win.

TABBY'S STAR

YURI MILNER

Entrepreneur, investor; physicist; founder, Digital Sky Technologies

Fifteen hundred light-years away, in the direction of Cygnus, lies a star that probably doesn't host an advanced civilization. You might think this is a non-story. In fact, it's big news.

In September, astronomers from the *Kepler* mission published a description of the star (officially designated KIC 8462852—unofficially, "Tabby's Star," after Tabetha Boyajian, the lead author of the paper).* It's around the same size as the Sun, but with a "bizarre" light curve—variations in the intensity of the light received from the star. A planet the size of Jupiter, passing in front of such a star, might be expected to dim the star's light about 1 percent. This particular star's light has been observed to drop 22 percent, in asymmetric and aperiodic dimming events unlike anything else seen by *Kepler*.

Possible explanations—none completely satisfactory—include a swarm of disintegrating comets or a disk of matter surrounding the star (which looks far too old to have retained such a disk). Jason Wright, an astronomer at Penn State University who was consulted by Boyajian about the problem, proposed an unlikely but intriguing possibility: that the dimming effect is caused by a swarm of Dyson Spheres—hypothetical megastructures that advanced civilizations might build to capture energy from their stars.

---

* http://arxiv.org/pdf/1509.03622v1.pdf

There are three reasons why all this is important.

First, *something* interesting is happening around Tabby's star. Even if it's not a megastructure, investigating it will surely increase our knowledge of stars, the formation of planets, or both.

Second, the anomaly was not discovered by astronomers. It was flagged by "citizen scientists" from the Planet Hunters project, scanning the *Kepler* data for signs of unknown extra-solar planets. This is a significant development in 21st-century science—its gradual broadening beyond academia, research institutions, and corporations to include the general public. Open data has allowed ordinary people to sample the immense harvest of data collected by instruments like the *Kepler* space telescope. Distributed computing enables them to use their personal computer power to analyze that data. And programs such as Planet Hunters invite them to use their critical faculties to find interesting patterns. We are witnessing the early steps of a revolution in the scientific process: the growth of a planetwide network of specialists, laypeople, and computers, collaborating to create scientific knowledge.

Third, the mere fact that astronomers can investigate a specific planet as a candidate for life illustrates how the *Kepler* mission has transformed astrobiology—from a heroic but marginal pursuit into a popular and rapidly maturing science. Less than a decade ago, many believed that potentially habitable planets were vanishingly rare. Today, to suggest that there are billions—in our galaxy alone—is a conservative estimate. And more and more evidence, such as the ubiquity of organic molecules in environments beyond our solar system, suggests that life may bloom on some of these planets.

Intelligent life, though, remains a great unknown. We know it has arisen once in at least 3.8 billion years of evolution on Earth. But extrapolating from Earth to the universe is guesswork.

Yet after *Kepler*, theories about civilizations beyond Earth are no longer stabs in the dark. Now, as Tabby's Star shows us, the scope for serious science has expanded enormously. Astronomers and committed nonscientists can study large and growing bodies of data for interesting patterns. When they find them, they can focus the wide resources of modern astronomy—from radio searches to optical spectroscopy to computational modeling—on individual candidate planets.

In this century, we finally have a serious chance of resolving Fermi's Paradox: *Where is everybody?*

There is no bigger question out there.

EXTRATERRESTRIALS DON'T LAND ON EARTH!

DAVID CHRISTIAN

Director, Big History Institute and Distinguished Professor in
History, Macquarie University, Sydney; author, *Maps of Time:
An Introduction to Big History*

Yesterday, no extraterrestrials landed! Or the day before! Or, despite many claims to the contrary, in any earlier period of human history. Or Earth history.

This is odd. There are several hundred billion stars in our galaxy and at least 100 billion galaxies in our universe. In the last twenty years, astronomers have detected lots of planets around nearby stars, so we know planets are common. In fact, there could be tens of billions, or even 100 billion Earth-like planets in our galaxy alone.

It's hard not to think that a lot of these Earth-like planets (a few million perhaps?) may have had histories a bit like our Earth. They may have spawned living organisms. On Earth, we have found life in many extreme environments, from deep-sea oceanic vents (where the current record-holder can survive at 120° C), to the inside of rocks, where they have to live very, very slowly in order to survive. Endospores can temporarily stop living (well, metabolism ceases) until things improve. Some bacteria may have jumped from Mars to Earth. So, life can exist in a wide range of environments, and today many astrobiologists believe that life might have existed on Mars and Venus and could exist even now on some of the moons of Jupiter and

Saturn, such as Io and Europa, which have lots of ice. All in all, it's beginning to seem that life of some kind could be common in the universe. It may be that the universe is quite bio-friendly.

If Simon Conway Morris and others are right and there is a limited number of pathways along which life can evolve, then any organisms that exist on other planetary systems may have evolved in ways not too dissimilar to the organisms on our Earth. Evolution may have converged on similar solutions in other star systems. Perhaps multicellular organisms have evolved many times. Perhaps many had ways of detecting light waves (eyes?), and perhaps many developed ways of computing or thinking (brains?). Our galaxy is 13 billion years old and most of its stars are older than our Sun, so most of its planetary systems should be older than Earth, which means they would have had much longer to evolve complex life-forms.

Here on Earth, life got going more or less as soon as our young planet was cool enough to have liquid water. That's fast, and hints that simple forms of life may be common. Four billion years later, large, intelligent creatures have appeared, and lots of them. One of those species crossed a critical threshold when it evolved such a powerful form of language that its members began to share their ideas and accumulate more and more information from generation to generation. As a result of its ability to learn collectively, that species (us) has built an astonishing store of knowledge, which enables us to control more and more of our environment, until now we dominate the planet. We have become a planet-changing species and we now live in what many scholars call the Anthropocene Epoch. We've even launched a few of us short distances into space and sent robots throughout our solar system.

On planets where evolution began millions of years earlier

than on Earth, you'd think evolution might have gone well past the crucial threshold of collective learning, past the production of a planet-changing species, and on, perhaps, to the point of colonizing nearby star systems. Could there be thousands of planets with species capable of collective learning? We can't know, but such an estimate is not impossible, and many of these planets could be orbiting the 4,500 star systems within 60 light-years of our Earth that make up our galactic neighborhood.

So where are the extraterrestrials? This was Fermi's famous question. The SETI program has been scanning the heavens for evidence of alien life since 1960. We haven't seen them. We haven't heard them either, or detected any other signs of their existence. Frank Drake, inventor of the Drake equation, which lists the factors we must take into account to estimate the likelihood of encountering other species like ourselves, thought that one of the crucial factors might be how long planet-changing species like us could survive.

And there's the rub. We're so clever that we've invented weapons that could ruin the biosphere in a few hours, and our energy-hungry civilizations seem to be degrading the biosphere and the climate systems on which we depend. Is it possible that planet-changing species like us never get past this stage? Do they all hit a wall when they reach their local Anthropocene? If so, such species may last for a few centuries or a millennium or two and then flicker out, perhaps after retreating to impoverished niches where they eke out a miserable existence before going extinct. That would mean that even if planet-changing species— species capable of telling stories and jokes, of painting and dancing, and building pyramids and spaceships—are common, they would all self-destruct. That would solve Fermi's problem!

Or perhaps some other planet-changing species actually

learned their lesson, maybe after a few self-inflicted catastrophes. Perhaps they decided not to aim too high, not to try to dominate their planet or their solar system or neighboring star systems but to live more sociably with their home planet and the other organisms that surrounded them, after realizing this was the only way of surviving. Perhaps we don't see them because, like Candide at the end of Voltaire's novel, they are all happily cultivating their own gardens. That would also solve Fermi's problem!

WE ARE NOT UNIQUE, BUT WE ARE VERY MUCH ALONE

ANDRIAN KREYE

Editor, *The Feuilleton* (Arts and Essays), *Sueddeutsche Zeitung*, Munich

It has been increasingly exciting to follow the recent surge in the discovery of exoplanets. Not only because what started as a needle-in-a-haystack endeavor in the late 1980s has become a booming field of space exploration, gaining momentum with the success of NASA's *Kepler* space telescope. As I write, the Exoplanets Data Explorer maintained by Jason Wright at Penn State lists 1,642 confirmed planets and 3,787 unconfirmed *Kepler* candidates.

There are severe downsides to most of those planets. Only 63 light-years away, for example, a blue-marble planet named HD 189733 b orbits its star. Daytime temperatures on this planet average 1,700 degrees Fahrenheit, wind speeds reach 7,000 mph, and the blue color in the atmosphere comes from rains of molten glass. Only four of the exoplanets found by now have the right distance from their stars to host life. This invites the conclusion that although our home planet is far from unique in the universe, we as humans are very much alone.

Most conclusions drawn from the discovery of exoplanets aren't quite as philosophical. Great findings about the history of the universe and the origins of life are made. While the glamour of space exploration lives on in the dreams of billionaire entrepreneurs and potentates, pop-culture ideas of settling space are gaining traction again. With the apocalyptic specter of climate change rendering this planet inhabitable, colonizing other planets seems an attractive idea.

Blockbusters like *Interstellar* and *The Martian* have used this longing for a life beyond our atmosphere for entertainment. But even when Harvard astronomer Dimitar Sasselov toured the lecture circuit a few years ago talking about the thrill of discovering faraway planets, you could sense the pangs of science-fiction longing in the audience. What if there indeed is life out there? Other habitable planets?

It is exactly those science-fiction dreams that fuel the news about the vast number of exoplanets of import. As symbols, they serve as extensions of the Blue Marble image of planet Earth taken by the Apollo 17 crew in December 1972. Back then, the Blue Marble showed us the reality of what Buckminster Fuller called "Spaceship Earth" just four years before—Earth being a rather small vehicle with finite resources. The Blue Marble went onto the cover of Stewart Brand's *Whole Earth Catalog*, the principal manual of the emerging ecological movement.

Even though the recent wave of anachronistic space-age glamour overshadows the great news about exoplanets, they still exemplify a shift in global consciousness. With all escape routes now officially closing (planet HD 189733 b being just one sensational example of the forbidding nature of space), the realization that humankind has to make the best of its home planet is taking hold, and not only in progressive circles. The recent climate talks in Paris have shown that the political will to take action finally transcends borders, ideologies, and national interests.

The symbolism of exoplanets goes beyond the Buckminster Fuller metaphor of Spaceship Earth. It shows that the drive of science knows no limits. When astronomers confirm the first extragalactic planets, the reach for infinity will open even wider realms of understanding of the universe. This understanding, coupled with a new consciousness about the value and fragility of our own planet, can foster a push for solutions right here on Earth—and strengthen the realization that the dream of habitable planets, or even communicating with alien life-forms, is as absurd as ideas about afterlives and deities.

BREAKTHROUGH LISTEN

MARTIN J. REES
Former president of the Royal Society; emeritus professor of
cosmology and astrophysics, Cambridge University;
author, *From Here to Infinity*

Searching for extraterrestrial intelligence (SETI) has for decades
been a fringe endeavor. But it's moving toward the mainstream.
In 2015, it gained a big boost from the launch of Breakthrough
Listen—a ten-year commitment by the Russian investor Yuri
Milner to scan the sky in a far more comprehensive and sus-
tained fashion than ever before.

It's a gamble: Even optimists rate the probability of success at only a few percent. And of course radio transmission
is only one channel whereby aliens might reveal themselves.
But the stakes are high. A manifestly artificial signal—even if
we couldn't decode it—would convey the momentous message
that intelligence had emerged elsewhere in the cosmos.

These searches are more strongly motivated than they were
in earlier decades. The *Kepler* space telescope, surely one of the
most cost-effective and inspirational projects in NASA's history,
has revealed that most stars in our galaxy are orbited by retinues
of planets. There are literally billions of them in the Milky Way
with the size and temperature of Earth.

But would these planets have developed biospheres? Or is
Earth unique and all others sterile and lifeless? Despite all we
know about life's evolution, its actual origin—the transition
from complex molecules to the first replicating and metabo-

lizing systems we would deem to be "alive"—has remained a mystery, relegated to the "too difficult" box. But it is now being addressed by top-ranking scientists. We may soon know whether life's emergence was a fluke or near-inevitable in the kind of "chemical soup" expected on any planet resembling the young Earth—and also whether the DNA/RNA basis of terrestrial life is special or just one of several possibilities.

In seeking other biospheres, clues will surely come from high-resolution spectra, using the James Webb Space Telescope, and the next generation of 30-plus-meter ground-based telescopes that will come online in the 2020s.

Conjectures about advanced alien life are of course far more shaky than those about simple life. We know, at least in outline, the evolutionary steps whereby nearly 4 billion years of Darwinian evolution led to the biosphere of which we humans are a part. But billions of years lie ahead. I would argue that our remote, posthuman descendants will not be "organic" or biological, and they will not remain on the planet where their biological precursors lived. And this offers clues to the planning of SETI searches.

Why is this? It's because posthuman evolution will be spearheaded by superintelligent (and supercapable) machines. There are chemical and metabolic limits to the size and processing power of "wet," organic brains. But no such limits constrain electronic computers (still less, perhaps, quantum computers). For these, the potential for further development could be as dramatic as the evolution from monocellular organisms to humans. So, by any definition of "thinking," the amount and intensity done by organic human-type brains will be utterly swamped by the cerebrations of AI. Moreover, the Earth's biosphere is not essential—indeed, it's far from an optimal environment—

for inorganic AI. Interplanetary space will be the preferred arena where robotic fabricators will have the grandest scope for construction, and where nonbiological "brains" may develop insights as far beyond our imaginings as string theory is for a mouse.

This scenario implies that even if life originated only on Earth, it need not remain a trivial feature of the cosmos: Humans may be closer to the beginning than to the end of a process whereby ever more complex intelligence spreads through the galaxy. But in that case there would, of course, be no "ET" at the present time.

Suppose, however, that there are other biospheres where life began and evolved along a track similar to what happened on Earth. Even then, it's highly unlikely that the key stages would be synchronized. A planet where it lagged significantly behind what has happened on Earth would plainly reveal no evidence of ET. But on a planet around a star older than the Sun, life could have had a head start of a billion years and already transitioned to the futuristic posthuman scenario.

The history of human technological civilization is measured in centuries—and it may be only one or two more centuries before humans are overtaken or transcended by inorganic intelligence, which will then persist and continue to evolve for billions of years. This suggests that if we were to detect ET, it would be far more likely to be inorganic. We would be most unlikely to "catch" it in the brief sliver of time when it took organic form. A generic feature of these scenarios is that organic human-level intelligence is just a brief prelude before the machines take over.

It makes sense to focus searches first on Earth-like planets orbiting long-lived stars (the "Look first under the lamppost" strategy). But science-fiction authors remind us that there are more exotic alternatives. In particular, the habit of referring to "alien civilizations" may be too anthropocentric—ET could be more like a single "mind."

Breakthrough Listen will carry out the world's deepest and broadest search for extraterrestrial technological life. The project involves using radio dishes at Green Bank, in West Virginia, and at Parkes, in New South Wales—and perhaps others, including the Arecibo Observatory in Puerto Rico—to search for non-natural radio transmissions, using advanced signal-processing equipment developed by a team based at UC Berkeley. More-over, the advent of social media and citizen science will enable a global community of enthusiasts to download data and partic-ipate in this cosmic quest.

Let's hope that Yuri Milner's private philanthropy will one day be supplemented by public funding. I'd guess that the mil-lions watching *Star Wars* would be happy if some of the tax rev-enues from that movie were hypothecated for SETI.

But in pursuing these searches we should remember two maxims, both oft quoted by Carl Sagan. First, "Extraordinary claims require extraordinary evidence," and second, "Absence of evidence isn't evidence of absence."

LIFE IN THE MILKY WAY

MARIO LIVIO
Astrophysicist; author, *Brilliant Blunders*

The question of whether extrasolar life (and extrasolar complex life in particular) exists, is arguably one of the most intriguing questions in science today.

While we don't know with any certainty whether the emergence of life on an extrasolar planet requires conditions similar to those on Earth, the presence of liquid water on a rocky surface is thought to be a generic necessity for life-producing chemistry to operate. This assumption has led to the concept of a habitable zone (HZ)—that "Goldilocks" not-too-hot, not-too-cold region around a star where the temperature and atmospheric pressure allow for liquid water to exist on the planet's surface. The idea of the HZ, in turn, has brought the question of how many Earth-size planets in the HZ exist in our Milky Way Galaxy to center stage.

Amazingly, during the past few years, observations (primarily by the *Kepler* space observatory) have accumulated sufficient statistics to solve this piece of the puzzle. Even conservative estimates, published in 2014, put the number of roughly Earth-size planets orbiting Sun-like stars in the HZ (in the Milky Way) at about 10 billion!

The publication of this empirically based estimate marked a critical point at which the quest for extrasolar life transitioned from mere speculation to an actual science. The realization that exoplanets could—in principle, at least—support life has turned the search for extrasolar life almost into an obsession for many astronomers. Plans for this field envisage a two-pronged attack:

1. A series of upcoming telescopes (in space and on the ground) will look for biosignatures—characteristics imprinted by life processes—in the atmospheres of planets in the HZ of their host stars.

2. Russian billionaire Yuri Milner announced in July 2015 a $100-million decadal project called "Breakthrough Listen" aimed at providing the most comprehensive search for alien communication (an extension of SETI) to date.

There is little doubt that the determination of the number of planets able to host life will stay news for at least a few decades. The only discovery in this domain that will eclipse these findings will be the actual *detection* of extrasolar life. We are, for the first time in human history, on the verge of potentially eliminating the last obstacle to Copernican modesty. We have discovered that neither our place in the galaxy nor our galaxy itself is special. Darwin has further shown that humans are a natural product of evolution by means of natural selection. The discovery of extrasolar life will demonstrate that even that last claim to being special will have to be abandoned.

THERE IS (ALREADY) LIFE ON MARS

MICHAEL I. NORTON

Harold M. Brierley Professor of Business Administration, Harvard Business School; co-author (with Elizabeth Dunn), *Happy Money: The Science of Smarter Spending*

Members of the Mashco-Piro tribe—previously viewed as one of the few remaining "uncontacted" peoples—have recently emerged to make increasing contact with the outside world. But this is a less than heartwarming story: As is so often the case with such contact, members of the tribe are vulnerable to unfamiliar diseases, such as influenza. These active efforts to become "contacted" create a problem for countries like Brazil, which have initiated far-sighted "no contact" policies to allow such tribes to choose seclusion. As José Carlos Meirelles, an indigene-protection agent in Brazil, put it: "If they are seeking contact, we must welcome them in the best manner possible. We must take care of their health, block out the boundaries of their territory, give them some time to adjust to the madness of our world."

Lately we've also learned more about the role of Catharine Conley, "planetary protection officer" at NASA, who has the job not of finding life on Mars but of protecting Mars from life on Earth. Scientists agree that life exists on Mars, if only in the form of microbes from Earth that took an accidental interplanetary ride. Despite the best efforts of NASA—which include sterilizing and sometimes even baking spacecraft—some life slips through. Much like the "no contact" policies for uncontacted peoples, NASA has protocols that keep rovers on Mars far from "special regions" where bacteria from Earth might thrive.

But what happens when life on Mars chooses not to wait? All of recorded history shows that life tends to find life—or more likely in this case, lichens tend to find lichens. Surely we should apply the lessons learned from centuries of genocide (accidental and intentional) of indigenous peoples on this planet to nascent forms of life on Mars. "If they are seeking contact, we must welcome them in the best manner possible. We must take care of their health, block out the boundaries of their territory, give them some time to adjust to the madness of our world."

THE BREATHTAKING FUTURE OF A CONNECTED WORLD

CHRIS J. ANDERSON

Curator, TED conferences, TED Talks; author, *TED Talks: The Official TED Guide to Public Speaking*

Our planet is growing itself a brain. That process is accelerating, and the project will determine the future of humans and many other species.

The major Internet and space technology corporations, among others, have confirmed multibillion-dollar investments to bring low-cost broadband Internet to every square meter of Earth's surface within ten years. They are building the railway tracks and freeways of the 21st century—but at global scale, and with breathtaking speed.

Five billion human minds are therefore about to come online, mostly via sub-$50 smartphones. And unlike the two billion who preceded them, their first experience of the Internet may not be clunky text but high-resolution video and a fast connection to whatever grabs their imagination.

This is a social experiment without historical precedent. Most of us built our Internet habits on top of years of exposure to newspapers, books, radio, and TV. Many of those soon to come online are currently illiterate. Who is going to win their attention and with what consequences? Local language versions of social media, Wikipedia, Porn? Video games? Marketing come-ons? Government propaganda? Addictive distractions?

Free education? Conversations with mentors in other countries empowered by realtime machine translation?

It's certainly possible to imagine a beautiful scenario in which, for the first time in history, every human can have free access to the world's greatest teachers in his or her own language; people discover the tools and ideas to escape poverty and bigotry; growing transparency forces better behavior from governments and corporations; the world starts to benefit from billions of new minds able to contribute to our shared future; global interconnection begins to trump tribal thinking.

But for all that to have even a chance of happening, we need to get ready to engage in the Mother of all attention wars. Every global company, every government, and every ideology has skin in this game. It could play out in many different ways, some of them ugly.

What's unique and significant is that we have a roadmap. It's now clear that we will not need to physically wire the planet. Satellites, possibly aided by drones and balloons, are about to get the job done a lot faster. We can therefore be certain that a massive transformation is about to hit. We'd better get ready.

EVERYTHING IS COMPUTATION

JOSCHA BACH
Cognitive scientist, MIT Media Lab/Harvard Program for
Evolutionary Dynamics

These days we see a tremendous number of significant scientific news stories, and it's hard to say which has the highest significance. Climate models indicate that we are past crucial tipping points and irrevocably headed for a new, difficult age for our civilization. Mark van Raamsdonk expands on the work of Brian Swingle and Juan Maldacena and demonstrates how we can abolish the idea of spacetime in favor of a discrete tensor network, thus opening the way for a unified theory of physics. Bruce Conklin, George Church, and others have given us CRISPR/Cas9, a technology that holds promise for simple and ubiquitous gene editing. "Deep learning" starts to tell us how hierarchies of interconnected feature detectors can autonomously form a model of the world, learn to solve problems, and recognize speech, images, and video.

It is perhaps equally important to notice where we lack progress: Sociology fails to teach us how societies work; philosophy seems to have become infertile; the economic sciences seem ill-equipped to inform our economic and fiscal policies; psychology does not encompass the logic of our psyche; and neuroscience tells us where things happen in the brain but largely not what they are.

In my view, the 20th century's most important addition to understanding the world is not positivist science, computer technology, spaceflight, or the foundational theories of physics. It is the notion of computation. Computation, at its core, and as informally described as possible, is simple: Every observation yields a set of discernible differences.

These we call information. If the observation corresponds to a system that can change its state, we can describe those state changes. If we identify regularity in those state changes, we are looking at a computational system. If the regularity is completely described, we call this system an algorithm. Once a system can perform conditional state transitions and revisit earlier states, it becomes almost impossible to stop it from performing arbitrary computation. In the infinite case—that is, if we allow it to make an unbounded number of state transitions and use unbounded storage for the states—it becomes a Turing machine, or a Lambda calculus, or a Post machine, or one of the many other mutually equivalent formalisms that capture universal computation.

Computational terms rephrase the idea of "causality," something that philosophers have struggled with for centuries. Causality is the transition from one state in a computational system to the next. They also replace the concept of "mechanism" in mechanistic, or naturalistic, philosophy. Computationalism is the new mechanism, and unlike its predecessor, it is not fraught with misleading intuitions of moving parts.

Computation is different from mathematics. Mathematics turns out to be the domain of formal languages and is mostly undecidable, which is just another word for saying "uncomputable" (since decision making and proving are alternative words for computation, too). All our explorations into mathematics are computational ones, though. To compute means to actually do all the work, to move from one state to the next.

Computation changes our idea of knowledge: Instead of

justified true belief, knowledge describes a local minimum in capturing regularities between observables. Knowledge is almost never static but progresses on a gradient through a state space of possible worldviews. We will no longer aspire to teach our children the truth, because, like us, they will never stop changing their minds. We will teach them how to productively change their minds, how to explore the never-ending land of insight.

A growing number of physicists understands that the universe is not mathematical but computational, and physics is in the business of finding an algorithm that can reproduce our observations. The switch from uncomputable mathematical notions (such as continuous space) makes progress possible. Climate science, molecular genetics, and AI are computational sciences. Sociology, psychology, and neuroscience are not: They still seem confused by the apparent dichotomy between mechanism (rigid moving parts) and the objects of their study. They are looking for social, behavioral, chemical, neural regularities, where they should be looking for computational ones.

Everything is computation.

IDENTIFYING THE PRINCIPLES, PERHAPS THE LAWS, OF INTELLIGENCE

PAMELA McCORDUCK
Author, *Machines Who Think; The Edge of Chaos; Bounded Rationality: A Novel; This Could Be Important;* co-author (with Edward Feigenbaum), *The Fifth Generation*

The most important news for me came in mid-2015, when three scientists—Samuel J. Gershman, Eric J. Horvitz, and Joshua B. Tenenbaum—published "Computational rationality: A converging paradigm for intelligence in brains, minds, and machines" in *Science*. They announced that something new was under way: an effort to identify the principles of intelligence, just as Newton once discovered the laws of motion.

Formerly, any commonalities among a stroll in the park, the turbulence of a river, the revolution of a carriage wheel, the trajectory of a cannon ball, or the paths of the planets seemed preposterous. It was Newton who found the underlying generalities that explained each of them (and so much more) at a fundamental level.

Now comes a similarly audacious pursuit, to subsume under general principles, perhaps even laws, the essence of intelligence wherever it's found. "Truth is ever to be found in simplicity, and not in the multiplicity and confusion of things," Newton said. As far as intelligence goes, we are pre-Newtonian. Commonalities of intelligence shared by cells, dolphins, plants, birds, robots, and humans seem, if not preposterous, at least far-fetched.

Yet rich exchanges among artificial intelligence, cognitive psychology, and the neurosciences, for a start, aim exactly toward Newton's "truth in simplicity," those underlying principles (maybe laws) that will connect these disparate entities together. The pursuit's formal name is computational rationality. What is it exactly, we ask? Who, or what, exhibits it?

The pursuit is inspired by the general agreement in the sciences of mind that intelligence arises not from the medium that embodies it—whether biological or electronic—but the way interactions among elements in the system are arranged. Intelligence begins when a system identifies a goal, learns (from a teacher, a training set, or an experience), and then moves on autonomously, adapting to a complex, changing environment. Another way of looking at this is that intelligent entities are networks, often hierarchies, of intelligent systems—humans certainly among the most complex but congeries of humans even more so.

The three scientists postulate that three core ideas characterize intelligence. First, intelligent agents have goals and form beliefs and plan actions that will best reach those goals. Second, calculating ideal best choices may be intractable for real-world problems, but rational algorithms can come close enough ("satisfice" in Herbert Simon's term) and incorporate the costs of computation. Third, these algorithms can be rationally adapted to the entity's specific needs—either offline, through engineering or evolutionary design, or online, through meta-reasoning mechanisms that select the best strategy on the spot for a given situation.

Though barely begun, the inquiry into computational rationality is already large and embraces multitudes. For example, biologists now talk easily about cognition, from the cellular to the symbolic level. Neuroscientists can identify computational strategies shared by both humans and animals. Dendrologists can show that trees communicate with each other (slowly) to warn of nearby enemies, like wood beetles: Activate the toxins, neighbor.

The humanities themselves are comfortably at home here too, although it's taken many years for most of us to see that. And of course here belongs artificial intelligence, a key illuminator, inspiration, and provocateur.

It's news now; it will stay news because it's so fundamental; its evolving revelations will help us see our world, our universe, in a completely new way. And for those atremble at the perils of superintelligent entities, surely understanding intelligence at this fundamental level is one of our best defenses.

NEURO-NEWS

NOGA ARIKHA

Historian of ideas; author, *Napoleon and the Rebel*

Science is never fixed in place: It must always move forward, and in that sense it is always "news." What makes science news in a journalistic sense, however, tends to be biased by current concerns, economic interests, and popular fears and hopes.

It is no surprise that research into the brain, in particular, continues to be the focus of much media attention—not only for the obvious reason of its central role in the very fabric of evolved life and of its infinite complexity but also because of the strong need to understand the biological bases of human behavior. This has led to many excessively positive claims for, and overinterpretations of, necessarily partial, provisional findings about brain mechanisms. The prefix "neuro" now twists into pseudoscientific shape all aspects of human behavior, from aesthetics to economics, as if the putative cerebral correlates for all that we do explained to us what we are. There are worthwhile and important avenues to explore here, but reports in the mainstream media hardly do justice to their scientific, methodological, and conceptual complexity.

Yet truly newsworthy neuroscience does get reported. The publication in a June 2015 issue of *Nature* of the discovery of a lymphatic system within the central nervous system is hugely important and was acknowledged as such in more mainstream venues. *ScienceDaily* titled its report of the discovery "Missing link found between brain, immune system; major disease implications," with the blurb, "In a stunning discovery that overturns decades of textbook teaching, researchers have determined that the brain is directly connected to the immune system by vessels previously thought not to exist. The discovery could have profound implications for diseases from autism to Alzheimer's to multiple sclerosis."

We might need to take with a few grains of salt this last sentence—the sort of claim that reflects wishful thinking rather than actual reality, typical of what constitutes fast-burning "news." On the other hand, few discoveries do "overturn decades of textbook teaching"—and this one probably does. The fact that established teaching can be overturned is important in itself; it's easy to forget that most work goes on within given frameworks on the basis of assumptions rather than with an eye to the need for questioning those assumptions. This particular discovery emphasizes at last the need to understand connections between the nervous and immune systems, and can only promote the development of the burgeoning field of neuroimmunology. Precisely because of the specialized nature of research and clinical care, brain facts tend to be understood apart from body facts, in a Cartesian fashion, as if one were really apart from the other. This piece of news reminds us that we can understand one only as an aspect of the other, and that we need to take seriously, in scientific terms, phenomena such as the placebo and nocebo effects and the role of the psyche in the evolution of mental and physical disease generally. And in turn, this shows that what we take to be scientific news—that is, news about our understanding of the world and ourselves—is a function of what we expect.

MICROBIAL ATTRACTIONS

PAMELA ROSENKRANZ
Artist

Sterility is not considered healthy anymore. Medicine is shifting from an antibiotic to a probiotic approach, and the idea of hygiene is becoming an organization of contamination, as opposed to disinfection. Not long ago, it was determined that the placenta is not sterile after all. The growing fetus was earlier believed to thrive in an absolutely clean bubble; instead, it seems to be confronted with germs through the filter of its mother's biological system and building its future immune system from the very start of cell division.

There are trillions of viruses, bacteria, fungi, and parasites thriving in each of us right now. Around 2 pounds of our body weight consists of what are popularly called bugs. Many of the microbiotic organisms are ancient. The most feared are viruses like Ebola or HIV, bacteria like *Streptoccocus*, and parasites like rabies. But next to the few fast and furious scary exceptions, most of the common organisms are easy for our immune system to deal with—even when they're pathogenic. New research suggests that many of them are actually keeping us healthy; they seem to be "training" our immune system.

The term "microbiome" was coined in the 1990s, but research is still beginning to sort the good from the bad. This community of organisms is so manifold and complex that we speak of a sea, a forest, a new natural world to be discovered within us. The main idea so far is that the more diversity—not

236

just in the environment we live in but also in the environment that lives within us—the better.

One simple clinical treatment that has turned into a substantial new industry is fecal transplants from healthy to sick people. It has been shown to heal the colon from an overgrowth of *Clostridium difficile*, a bacteria that often cannot be cured by antibiotics. It has helped obese people lose weight. And as it turns out that the gut is fundamentally intertwined with our brain, it influences our psychological sanity.

Current research suggests that certain bacterial cultures cause anxiety, depression, and even Alzheimer's, while others might help alleviate these ailments. But the effects on our mental state seem even more shockingly direct, if we take toxoplasmosis as an example: This neuroactive parasite influences one of the most existential of our feelings—sexual attraction.

We, along with mice and other mammals, are only intermediary hosts; cats are its main target. In this unconscious *ménage a trois,* the parasite wants the mouse to be attracted to the cat, so *Toxoplasma gondii* travels up to the region of the mouse's brain where sexual arousal occurs and there it prompts the mouse to react positively rather than negatively to cat pheromones and approach the cat instead of fleeing—so that the cat can much more easily catch and ingest it. Once inside that cat, the parasite has reached its goal: It can reproduce.

Humans are part of its scheme in more abstract ways. Those who carry it are attracted to scents that originate from cat pheromones (this scent can be found in many perfumes—allegedly, Chanel No. 5, for instance). About 30 percent of the global population is infected—quite a target group! Apparently, this segment of humanity is also more prone to involvement in car accidents, and female carriers are known to acquire more designer clothes.

We tend to see sexuality as one of the main markers of our individuality, but not only does our own biological system react to sexual attractions in ways we can't control but there are also parasites that can neurologically influence, or possibly even direct, our behavior. It's a provocative and difficult topic, and it challenges the fundamental understanding of what it means to be human.

We are in constant exchange with our germs. We shake hands, kiss, have sex, travel, use toilets, go to parties, churches. Now there's research about how religion, as a social factor, might be entangled in that complicated communal sharing of microbial organisms. When we come together, what do we really exchange? Might it be that our need for social interaction is also influenced by the secret powers of microbes?

THE EPIDEMIC OF ABSENCE

MATT RIDLEY

Science writer; Fellow, Royal Society of Literature and the Academy
of Medical Sciences; author, *The Evolution of Everything*

As Stewart Brand acutely says, most of the things that dominate
the news are not really new: Love, scandal, crime, and war come
round again and again. Only science and invention deliver truly
new stuff, like double helixes and search engines. In this respect,
the new news from science that most intrigues me is that we
may have a way to explain why certain diseases are getting worse
as we get richer. We are defeating infectious diseases, slowing or
managing many diseases of aging, like heart disease and cancer,
but we are faced with a growing epidemic of allergy, autoimmu-
nity, and things like autism. Some of it is due to more diagnosis,
some of it is no doubt hypochondria, but there does seem to be
a real increase in these kinds of problems.

Take hay fever. It is plainly a modern disease, far more
common in urban, middle-class people than it used to be in
peasants or still is in subsistence farmers in Africa. There's good
timeline data on this, chronicling the appearance of allergies as
civilization advances, province by province or village by village.
And there's good evidence that what causes this is the suppres-
sion of parasites. You can see this happen in Eastern Europe and
in Africa in real time: Get rid of worms and a few years later children start getting hay fever. Moises Velasquez-Manoff chronicles
this in glorious detail in his fine book *An Epidemic of Absence*.

This makes perfect sense. In the arms race with parasites,

immune systems evolved to "expect" to be down-regulated by parasites, so they overreact in their absence. A good balance is reached when parasites try down-regulating the immune system, but it turns rogue when there are no parasites. And the obvious remedy works: Ingest worms and you rid yourself of hay fever. Though it's probably not worth it—worms are no fun.

But how many of our modern diseases are caused by this problem, an impoverished ecology not just of parasites but of commensal and symbiotic microorganisms too? Do kids today in the rich world have unbalanced gut flora after an upbringing of obsessive hygiene? Probably. How many diseases and disorders are the consequence of this? More than we think, I suspect—multiple sclerosis, obesity, anorexia, perhaps even autism.

There's a fascinating recent study by Jeffrey Gordon's group at Washington University School of Medicine, St. Louis, showing that if you take the gut flora from an obese person and introduce it into a mouse with no gut flora, the mouse puts on weight faster than does another mouse with gut flora introduced from the obese person's non-obese twin. That's a well-designed experiment.*

So a big new thing in science is that we are beginning to understand the epidemic of absence.

---

*   *Nature* 444: 1027–1031 (21 December 2006) | doi:10.1038/nature05414

BUGS R US

NINA JABLONSKI

Evan Pugh University Professor of Anthropology, Penn State University; author, *Living Color: The Biological and Social Meaning of Skin Color*

Ignaz Semmelweis changed our world when, back in 1847, he decided to start washing his hands after he performed an autopsy and before he delivered a baby. When Semmelweis worked in a Viennese obstetric hospital, the germ theory of disease and the concept of "infection" were unknown. Postpartum infections due to "childbed fever" killed a high percentage of women who gave birth in hospitals. Semmelweis knew that there was something in and around dead bodies that had the potential to cause disease, and so he decided to follow the practice of midwives and wash his hands before delivering a baby. Fewer mothers died, and Semmelweis knew he was onto something. During his lifetime, his innovation was rejected by fellow male physicians, but within decades, evidence from doctors and scientists in other parts of Europe proved him right. Small organisms like bacteria caused disease, and taking simple precautions like hand-washing could lower disease risk.

Thanks to Semmelweis and his intellectual descendants, we follow a range of routines from boiling water and avoiding tropical ice cubes to near-fanatical levels of hand sanitizing, in order to reduce the chances of getting sick because of the nasty bugs in our environment.

We have known for a long time that our bodies harbor lots

of "normal" flora, but until about a decade ago few people studied them. We focused on Semmelweis's disease-causing bacteria, which we cultured in petri dishes so that we could identify and kill them. The rest of our microbial residents were thought to be pretty much harmless baggage and were ignored.

The introduction of new methods of identifying diverse communities of organisms from their DNA alone (including such innovations as high-throughput DNA sequencing) changed all that, and we began to realize the magnitude of what we had been missing. The world of critters living in and on us was soon discovered to be a vast and complex one, *and it mattered*.

Since 2008, when the Human Microbiome Project officially started, hundreds of collaborating scientists have illuminated the nature and effects of the billions of bacteria that are part of our normal healthy bodies. There isn't one human microbiome; there are many. There is a microbiome in our hair, one up our nostrils, another in our vaginas, several lavishly differentiated on the vast real estate of our skin, and a veritable treasure trove in our gut, thanks to diligent subcontractors in the esophagus, stomach, and colon.

This great menagerie undergoes changes as we age, so that some of the bacteria that were common and apparently harmless when we were young start to bother us when we're old, and *vice versa*. The taxonomic diversity and census of our resident bacteria are more than just subjects of scientific curiosity; they matter greatly to our health. The normal bacteria on our skin, for instance, are essential to maintaining the integrity of the skin's barrier functions. Many diseases—psoriasis, obesity, inflammatory bowel disease, some cancers, even cardiovascular disease—are associated with shifts in our microbiota.

While it's too early to tell if the changing bacteria are the cause or the result of these afflictions, the discovery of robust associations between bacterial profiles and disease states opens the door for new treatments and targeted preventive measures. The body's microbiota also affect and are affected by the body's epigenome, the chemical factors influencing gene expression. Thus, the bugs on us and in us are controlling the normal action of genes in the cells of our bodies, and changes in the proportions or overall numbers of bacteria affect how our cells work and respond to stress.

Let's stop thinking about our bodies as temples of sinew and cerebrum, and instead as evolving and sloshing ecosystems full of bacteria that are regulating our health in more ways than we could ever imagine. As we learn more about our single-celled companions in the coming years, we'll take probiotics for curing acute and chronic diseases, we'll undertake affirmative action to maintain diversity of our gut microflora as we age, and we'll receive prescriptions for increasingly narrow-spectrum antibiotics to exterminate only the nastiest of the nasties when we have a serious acute infection. Hand sanitizers and colon cleansing will probably be with us for some time, but it's best just to get used to it now: Bugs R us.

FECAL MICROBIOTA TRANSPLANTS

JOICHI ITO
Director, MIT Media Lab

Although we have been talking about the microbiome for years, news about our microbial friends was huge this year. We have known for some time that the microbes in our gut were import- ant for our health, but recently studies are beginning to show that the gut biome is even more important than we thought.

Fecal Microbiota Transplantations, or FMTs, have been shown to cure *Clostridium difficile* infections in 90 percent of cases, a condition notoriously difficult to treat any other way. We don't know exactly how FMTs work, other than that the introduction of microbiota (poop) from a healthy individual somehow causes the gut of an afflicted patient to regain its mi- crobial diversity and rein in the rampant bacteria.

It appears that our gut microbes produce a wide variety of neurotransmitters that influence our brains, and vice versa, much more than previously believed. There is evidence that in addition to mood, a number of brain disorders may be caused by microbial imbalance. The evidence is so strong that FMT banks, such as OpenBiome, have started screening donors for psychiat- ric problems along with a variety of health issues. Consequently, it is now harder to qualify as a fecal-bank donor than to get into MIT or Harvard. Perhaps machines can help us here, as they do everywhere else; Robogut is making headway in creating synthetic poop.

It has been shown that mice without gut microbes socialize less than mice with proper gut biomes, causing scientists to theorize that while socialization doesn't help the fitness of mice, their social behavior and their habit of eating each other's feces may be driven by the microbes "wanting" to be shared among the mice.

Many of our favorite foods are really the favorite foods of our gut microbes, which turn those foods into things our bodies need and like. Also, it appears that oligosaccharides—abundant in breast milk and regarded as metabolically "inert"—selectively feed some of our "good" gut microbes. Not only are microbes more abundant in the human body than human cells, it seems they may be the reason we do many of the things we do. They may well be as important, if not more important, in many of our body's processes than our own cells are.

However, not all microbes are good for us. In fact, most are neutral and some are bad for us. Take, for example, *Toxoplasma gondii,* which causes infected rats to lose their natural fear of cats because this parasitic protozoan needs to get inside cats to reproduce. Or rabies, which causes animals to attack other animals to increase transmission.

And the microbes are everywhere. The detergents we use have eliminated the ammonia-oxidizing bacteria (AOB) on our skin—bacteria present on the skin of the Yanomami, indigenes of the Amazon rain forest. Pre-modern-hygiene tribes like the Yanomami do not suffer from acne or most forms of inflammatory skin diseases. In a study of more than 1,000 Kitavan islanders of Papua New Guinea, there was not a single such case. There is increasing evidence that allergies and many other modern ailments arose only after the invention of modern hygiene.

The microbes in the air are also part of the system. Studies show that infection rates in hospitals decrease if you open a window and let the diverse outdoor microbes in, compared with such rates in hospital environments that filter and sterilize the air. Microbes in the soil are an essential part of the nutrient system for plants, and the microbes in the plants allow the plants to convert the nutrients into flavors and nutrients for us. Using artificial fertilizers that destroy the microbial flora of our soil and "enriching" our blank calories with oversimplistic vitamins that happen to be the *molecule du jour* are doubtless the opposite of what we should be doing.

The human gut, particularly the colon, has the highest recorded microbial density of any known microbial habitat. Our gut is almost the perfect environment to support the biodiversity and complexity that is our gut biome. The temperatures are well regulated, with us, the human hosts, able to survive in extreme conditions and sharing microbes with these varied environments. From the perspective of the microbes, we are an almost perfectly evolved life-support system. It may be arrogant to think of the microbes as some sort of "little helpers;" perhaps it's more accurate to think of ourselves as architectural innovations created by the microbes.

HI, GUYS

ALAN ALDA

Actor; writer; director; Visiting Professor, Stony Brook University School of Journalism; author, *Things I Overheard While Talking to Myself*

This year I had the wonderful and shocking awareness that I'm not only connected to microbes but, in a way, I'm so dependent on them that I sort of *am* them.

Darwin gave me the understanding that I'm related to the rest of the beasts of the Earth, but work on the microbiome, released in 2015, impressed me with how much a part of me microbes really are and how much I look to them for my very existence.

It started with the spooky information a short while back that there are ten times as many of them in me as there are me in me—at least if you compare the number of their cells in me to the number of mine in me.

From what I read, they're so specialized that the microbes in the crook of my arm are more like the ones in the crook of *your* arm than they are like the microbes in my own hand.

Then came the discovery that before long I'll be able to get a fecal transplant, or maybe simply take a poo pill, to relieve all kinds of disturbances in my body—possibly even obesity, should it ever win the war against my self-control.

And there was the equally strange news that I give off a cloud of microbes wherever I go—and if they settle on a surface, someone could take a reading and record a kind of fingerprint of my personal microbiome after I'd left the scene.

They're ubiquitous little guys. There are, I believe, more of them pound for pound than any other living thing on Earth, and we can't even see them.

And they're powerful. One kind of microbe expands when wet, and a pound or two of them could lift a car a couple of feet off the ground. You could change your tire with them.

We've planted a flag on a new New World. The last frontier has just changed again, from outer space, to the brain, to an invisible world without which there would be no world as we know it.

Hi, guys.

THE ANTI-DEMOCRATIC TREND

DIRK HELBING

Professor of Computational Social Science, ETH Zurich

The digital revolution progresses at full pace and reshapes our societies. Many countries have invested in data-driven governance. The common idea is that more data is more knowledge, more knowledge is more power, and more power is more success. This magic formula has promoted the concept of a digitally empowered benevolent dictator, or "wise king," able to predict and control the world in an optimal way. This seems the main reason for the massive collection of personal data, which companies and governments alike have engaged in.

The concept of the benevolent dictator implies that democracy would be overhauled. I agree that democracy deserves a digital upgrade, but in recent years many voices in the IT industry have claimed that democracy is outdated and needs to be bulldozed. Similar arguments have come from politicians in various countries. Democracy is now in acute danger of ending, in response to challenges such as climate change, resource shortages, and terrorism. A number of countries come to mind.

However, recent data-driven analyses show that democracy is not a luxury. Rather, it pays off. A 2015 study by Heinrich Nax and Anke Schorr, using high-performance computers, reveals that "the growth of countries that democratize is generally faster and more sustained. The only exceptions are short-term incentives to de-democratize for the richest and most democratic countries, but such de-democratizations come with reduced growth beyond the short-run."* In other words, demolishing democracy would be a costly mistake.

The anti-democratic trend in many countries is dangerous and needs to be stopped. First, because it would further sociopolitical instability and end in revolution or war. (Similar instabilities have already occurred, in the transition from agricultural to industrial society and from that to the service society.) Second, because the magic formula noted above is based on flawed assumptions.

Society is not a machine. It cannot be steered, like a car. Interaction—and the resulting complex dynamics of the system—changes everything. We know this, for example, from spontaneous breakdowns of traffic flow. Even if we could read the minds of all drivers, such phantom traffic jams could not be prevented. But there is a way to prevent them, using suitable driver-assistant systems: distributed-control approaches, Internet of Things technology, realtime data, and suitable realtime feedback, together with knowledge gained from complexity science.

The paradigm of data-driven optimization might work if we knew the right goal function; moreover, the world would have to change slowly enough and things would have to be sufficiently simple and predictable. These conditions are not being fulfilled. As we continue to network the world, its complexity grows faster than the data volume, outstripping the processing power and the data that can be transmitted. Many aspects of our world are emergent and hardly predictable. Innovation is burgeoning and we need even more of it! Not even the goal function is obvious: Should it be GNP per capita or sustainability, power or peace, average life span or happiness? In such

---

* http://papers.ssrn.com/sol3/papers.cfm?abstract_id=2698287

cases, (co-)evolution, adaptation, and resilience are the right paradigms, not optimization.

Decision makers around the globe must recognize the need to preserve democracy and replace information systems based on mass surveillance and brute-force data mining. They need to argue for interdisciplinary and global collaboration, for approaches built on transparency and trust, for open and participatory systems, because those mobilize the capacity of an entire society. They need to promote systems based on diversity and pluralism to foster innovation, societal resilience, and collective intelligence.

If we don't manage to get things right, we may lose many societal, economic, legal and cultural achievements of the past centuries; we might see one of the darkest periods of human history, something much worse than *1984*'s "Big Brother is Watching You"—a society in which we lose our freedom, enslaved by a "citizen score" that gives us plus or minus points for everything we do, where governments and big corporations determine how we should live our lives.

THE AGE OF AWARENESS

QUENTIN HARDY

Deputy technology editor, *New York Times*

We are entering the Age of Awareness, marked by machine intelligence everywhere. It is a world instrumented with sensors that constantly describe the location and state of billions of people and objects, transmitting, analyzing, and sharing this information in cloud computing systems that span the globe. We are aware of innumerable interactions and increasingly capable of statistically projecting outcomes.

The scientific breakthroughs will depend not just on these tools but equally on the system into which they are integrated. The biggest changes and breakthroughs from the instrumented world bring together once disparate sectors of computing, which, by working in unison, create new approaches to product design, learning, and work.

The sectors include mobility, sensors, cloud computing, and data analysis, whether by machine learning or artificial intelligence. Sensors don't just give us new information about nature and society, they inform the configuration of cloud systems; the behavior of the analysis algorithms is likewise affected by the success of the changes they make to the cloud system, the sensors, and the external environment.

The result is a kind of flywheel world, in which data that were once stored and fetched now operate in streams, perpetually informing, changing, and being changed. The accelerating rate of change and increasing pace of discovery is a result of this shift. On a pragmatic level, it means that we will design much of the world to be in a potential state, not a fixed one. The focus of economic value is on changes that continually result from these interactions. Another outcome of this world of continual response and adaptation is the end of the 2,500-year-old (and increasingly suspect) Aristotelian project of creating a state of final knowledge. Instead, we truly live in change, pursuing the best optimization of knowledge.

Inside the flywheel world, the eternal present of consciousness within a solitary self is being modified by a highly connected and global data storage of the past, computation of the present, and statistical projection of the future.

We already see our human habits changing with the new technology, much the way print once re-oriented political and religious consciousness, or society changed to suit industrial patterns. As people, we are starting to imitate a software-intensive cloud computing system. Billions of people are gaining near-infinite abilities to communicate across languages to billions of other people. Artificial intelligence agents within those systems will track people, teaching and assisting them, and to yet-unknown extents reporting on the individuals to corporate (and possibly government) masters.

Learning is increasingly a function of microcourses that teach what you need to know and (thanks to analysis in the system) what you need to know next. We perceive life's genetic code as an information system, and we are learning how to manipulate it, either to hack the human body or to use DNA for unimaginably small and powerful computers that could extend greatly our powers of awareness and control.

Unique among times when technology has changed worldviews, this Age of Awareness knows that it is remaking the consciousness and expectations of being human. Gutenberg in 1450, or an industrialist in 1810, had no awareness of an effect on humans wrought by new technologies. Everyone now building the instrumented, self-aware planet can see and analyze the effects of their labor. That does not, to date, significantly improve our ability to plan or control its outcomes.

A LARGE-SCALE PERSONALITY RESEARCH METHOD

NATHALIE NAHAI

Web psychologist; author, *Webs of Influence: The Psychology of Online Persuasion*

The most important news of 2015 for me came in June, with the publication of "Automatic Personality Assessment Through Social Media Language" in the *Journal of Personality and Social Psychology*. For those working at the intersection of psychology and technology, the results of this study confirmed what many of us had been anticipating: the validation of a cheap, naturalistic, large-scale research method designed to assess and interpret the linguistic interactions that millions of us engage in online, every single day.

With a sample of over 66,000 active social-media participants, the researchers used a rich, open-vocabulary approach to build a predictive model of personality, using the "Big Five" personality traits of openness, conscientiousness, extraversion, agreeableness, and neuroticism. The methodology they employed yielded more accurate language-based predictions of personality than any other study to date, demonstrating not only a robust alternative to existing approaches but also that this kind of research can now be accomplished on an unprecedented scale and level of accuracy.

General insights into a population's personalities may not seem particularly consequential. We might know, for example, that individuals who score highly for extraversion prefer using more positive emotional words (such as "amazing," "great," "happy"), whereas those who score higher in neuroticism tend to use first-person singulars (such as I, me, mine) with greater frequency. But it's not until we get multiple data points at scale that a more profound picture emerges.

Considering the ease with which we can create unique profiles for users with little more than a few cookies and an IP address, we are now in the unique position of being able to cluster traits together and compile overall personality dispositions for millions of users, which can then be stored in psychometric databases. In fact, several companies have already begun this task, with commercial applications in mind.

Given that certain personality dispositions are associated with a range of predictable life outcomes (for instance, a propensity to risk-taking behaviors within high-scoring extravert populations), it's conceivable that such data could be used to affect the quality of our lives for good and for bad. This is where the importance of the research kicks in.

On the positive side: If we can design programs that make predictions about our personality by assessing publicly available data (our written interactions across social-media channels), this may prompt us to discover more about our motivations, our behaviors, and ourselves. It may also lead to smarter advertising and applications that will better serve our needs.

On the negative side: Outside the realm of academic research, such data-mining practices do not yet require consent and could therefore be used by any entity able to profile and categorize people (whether as citizens, customers, or potential employees) without their knowledge and beyond their control. Such information could then be used to determine whether to grant certain people access to particular services (such as lines of credit or medical insurance), career paths, and even citizenship.

Given the predictive potential of such a system, it is of vital importance that this news enter the public discourse, so that we are all better equipped to understand how the information we share online may be used to reveal potentially intimate aspects of ourselves. Only then can we make an informed choice as to how (or whether) to engage online.

THE CONQUEST OF HUMAN SCALE

CHARLES SEIFE

Professor of journalism, New York University; former staff writer, *Science*; author, *Virtual Unreality*

It was just one among dozens and dozens of revelations about the National Security Agency, barely enough to cause a stir in the papers. Yet it is a herald of a new era.

In May 2014, journalists revealed that the NSA was recording and archiving every single cell-phone conversation that took place in the Bahamas. Now, the Bahamas isn't a very big place, with only a few hundred thousand people on the islands at any given time. Nor is it often the source of international headlines. So, at first glance, the NSA's achievement might not look like much. But in capturing—and storing—all of the Bahamas' cell-phone conversations in real time, the NSA has managed to transform a significant proportion of the day-to-day interactions of a society into data—into information that can be analyzed and transformed and correlated and used to understand the people who produced it. This was something unthinkable even a decade or two ago, yet almost unnoticed, the processing power of computers, the scale of their memory banks, and the cheapness and ubiquity of their sensors are making civilization-scale data-gathering almost routine.

The NSA's collection program in the Bahamas—code-named SOMALGET—was a small part of a larger operation, which itself is just a tiny fraction of the NSA's global surveillance system. Whistleblowers and leaked documents have revealed that the NSA has been gathering and storing emails, phone calls, and other records on a global scale, and, apparently to capture entire nations' communications outputs and store them for later study. And the NSA isn't the only entity with such ambition. Other agencies and companies around the world have been collecting and creating data sets that capture one entire facet of the behavior of millions or even billions of people. The city of New York can now analyze each taxi ride taken in the five boroughs over the past several years. Google has stored every single character that anyone has entered into its search engine for more than a decade. It's all in there, taking up much less room in memory than you might think.

A medical researcher can now download and analyze all the drug prescriptions filled in the United Kingdom; an epidemiologist can view all the deaths recorded in the United States; a civil engineer can view all airline flights taken anywhere in the world at any time in recent history. Personal genomics companies are now performing cut-rate genomic analysis of more than a million customers; at this point, it's just cost and inclination that keeps us from capturing the genome of every individual on Earth. And as digital cameras, microphones, and other sensors are woven into every aspect of the fabric of our society, we aren't far from being able to capture the movements and utterances of every single macroscopic creature in the places we inhabit.

Pretty much anything that can be digitized or digitally collected and numbers in the billions or trillions or quadrillions can now be archived and analyzed. All our communications, our purchases, our travels, and our daily routines are to at least some degree sitting on banks of computer memory. We no longer have to guess, to sample, to model; it's all there for the taking. As this data begins to shine light into every corner of our society, we will recognize how much of our existence has been in darkness—and how different life will be in a world without shade.

BIG DATA AND BETTER GOVERNMENT

MARGARET LEVI
Jere L. Bacharach Professor of International Studies, University
of Washington; director, CHAOS (Comparative and Historical
Analysis of Organizations and States) Center; co-author (with
John S. Ahlquist), *In the Interest of Others*

Big Data gives business, government, and social scientists access
to information never available before. With the right tools of
analysis—which are improving exponentially as I write—Big
Data will transform the way we understand the world and the
means we use to fix problems. The U.S. and other governments
are building the capacity to use Big Data as a basis for determining best practices; university-based research programs are generating appropriate analytic tools; and various nonprofits around
the world are linking technology, data, and citizens to enhance
the implementation of government programs and services.

Science can now effectively be brought to bear on public
policymaking. Yet, important distinctions exist among the key
players. One set of actors wants to ensure that public policies
are evidence-based and a second set aims to enable citizens to
complain about poor services and to get the services they need.
Some are fundamentally concerned with the science and others
with voice.

Evidence-based policy has become a mantra in some circles, and increasingly the focus is on assessment of policies
once enacted as well as on the ex-ante crafting of good policy.

Randomized experiments have gained popularity worldwide by bringing scientific rigor into the appraisal of interventions meant to improve well-being. But they are not the only tools in the toolbox. Observational analyses using Big Data are just as important, particularly where randomization of people and communities is undesirable, infeasible, unethical, inadequate, or all of the above. Political considerations often trump randomization when it comes to the location of hospital facilities, military bases, and schools. Even in politicized circumstances, new techniques of causal inference from observational data make it possible to learn about the conditions under which different policies are likely to succeed. Indeed, the progress in recent years on generating scientific inferences from observational data has been breathtaking.

Simultaneously, another group of actors are stepping up to the plate to adapt and improve current technologies, data platforms, and analytic advances in the service of citizen voice. Providing individuals with mobile phones to take pictures, send texts and emails, and otherwise document what they see offers citizens a means for reporting on where things are broken and demanding that they be fixed. It is also a new and important form of quality control over elections, services, and bureaucrats. Reporting leaking gas mains or water hydrants, photographing potholes and abandoned homes, and naming corrupt officials can lead to significantly improved government responsiveness— and in some places already has, generally as a result of the work of nonprofits such as Code for America in the U.S. and eGovernments Foundation in India, or of university-based research teams collecting evidence on how government actually functions. One recent success involves discovering and correcting the gap in the distribution and use of food stamps in California.

The amount and kind of data collected from all of us does pose dangers to privacy and misuse. Science and engineering are being mobilized to ensure that the proper protections are in place, but governments must also convince publics that they are trustworthy in how they use the data they access. At stake is the promise of better government that draws on scientific analysis of policy and scientific and technological amplifiers of voice.

THIS IS THE SCIENCE-NEWS ESSAY YOU WANT TO READ

MARTI HEARST

Computer scientist, UC Berkeley School of Information; author, *Search User Interfaces*

Scientists and engineers continue to make progress in the battle against the overload of confusing choices that plague modern society. In response to well-known studies from the 1990s and 2000s which found that when presented with too many choices, people often opt to choose nothing at all, efforts in both the research and commercial worlds focused on mining behavioral "Big Data."

Now intelligent systems can predict what people want before they want it, so instead of offering a choice of navigational options on a Web site and forcing the consumer to choose among them, the smart app simply shows the two or three choices that are just right for that person. And instead of scanning the news presented by reputable news outlets, readers are shown just the right article, personalized for them, so they don't have to think about how they'll stay up-to-date. Just the movie or video you want to watch at this moment appears before your eyes as you settle into your chair. You don't have to give it a second thought! And of course your voting choices are arrayed for you in your favorite color scheme.

And it doesn't stop with reading. Your vacation planning is figured out for you now as well. In the past, before Intelligent Planning, you would never have thought your dream location was a small town in Kansas, but that is indeed your top recommendation, and so of course that's where you and your loved ones will have the best time. This way, the people who designed the system won't feel crowded in their vacation spots in Kauai.

So the science news is all good, except for the anti-science Huxley protesters who had contrary thoughts, but that information was not in the essay you wanted to read. *This* was the science-news essay you wanted to read (based on essays you've recently read, thoughts you've recently had, the *el grande* burrito currently in your intestinal tract, and the promotion you did not get last month at work).

And this is the science-news essay I wanted to write. (AI-Generator™)

Gentle reminder: Contrary thoughts experienced during the consumption of this essay will be reported.

THOSE ANNOYING ADS? THE HARBINGER OF GOOD THINGS TO COME

ROGER SCHANK

Cognitive psychologist; founder, Socratic Arts and XTOL; executive director, Engines for Education; author, *Education Outrage*

The most important news relevant to our future lives in the world of today's technology and is not exactly news. In fact, it's quite annoying. We all hate it. I am referring to those ads that pop up while you're doing something on the Internet, when you least want to see them.

The annoyance with those is news every day it seems. So here is the interesting question: Why might they be a good thing?

First let's discuss why this nuisance happens. Ads target you because of what you're doing on the Internet. For a while, I got ads for online nursing schools, because I had checked on an online nursing school to see what it was doing (because of my interest in online education, not nursing). If a computer can even come close to figuring out your interests, expect a targeted ad. Looking at suitcases online? You will soon receive suitcase ads. Now, this seems rather stupid and it's usually annoying. But it does work sometimes, so it will keep happening.

We're in the keyword stage of advertising. We are being told that this is science: IBM's Watson is doing deep learning. Don't be fooled. It's all keyword search and there's no science behind it. Directed advertising is all about keywords. Anything you type online is being tracked, by a machine that can count. No science is going on.

So what is the good news?

Having someone (or something) track you might not be such a bad thing. We like it when a map program knows where we are and we can figure out how to get where we're going. Many people like hook-up sites that tell you who's nearby whom you might like. But, here again, no science. There could be science. Hook-up sites might figure out whom you might like and tell you what you have in common to discuss. Will this happen? We're not that far away from it. We'd need a computer that knew about you the way a friend does (as opposed to your Web-surfing habits).

Now let's take this idea one step further. Suppose you were trying to fix a device in your home and that the device knew you were doing so and offered help along the way. That wouldn't be so bad. Suppose you were cooking something and the cookbook knew what you were cooking, what ingredients and equipment you had around, and could help you cook, modifying its recipes as needed. Suppose it saw you were doing it wrong and offered help? To do that, we need a model of your goals and the things that make you happy (and maybe a little physics).

Pushing the smart-machine idea even further, we can well imagine that if you were driving somewhere with a friend, the friend might say: "Hey, isn't that restaurant you like so much near here? Why not stop for a bite?" Is that an annoying ad or helpful advice? It depends on the situation and who said it, I suppose.

Let's move on to something more serious. My stomach hurts. I tell this to my wife and she suggests a medicine in the cabinet that she remembers I have used before and reminds me that it helped. Now, suppose this was not my wife but a computer? Is it an ad? Does it matter? Can we do this? Yes. AI technology could easily employ models of people and their needs. (But today we're busy with keywords.)

Imagine I am really sick. I'm afraid I'm having a heart attack. Today we could go to the ER, or more likely we search "heart attack symptoms." Maybe we call a doctor we know (assuming we know one who will answer right away). But in the future, the best and brightest cardiologists will be a click away, ready to answer your questions, offer suggestions, and maybe tell a few stories they're reminded of by situations like yours. Is this possible? Indeed it is. It requires indexing stories the way people do to get reminded. We have programs that do this already. But, sad to say, this is not on the agenda of commercial entities in AI just yet.

Very soon, AI programs will be good enough, not because they analyze keywords or do "deep learning" but because they can model situations and match them to what people have said about those situations. Imagine a video database of hundreds of thousands of experts. Well, "How would I search through all those stories?" is the natural question. We ask that question because searching is an everyday activity now, and it has taught us to believe in search, and everyone selling AI espouses the usefulness of keywords.

But it is not keywords that will cause this breakthrough. There's too much information to search through, and often what we need isn't there in the first place. But this isn't a search problem, it's a problem not unlike the getting-the-right-ad-to-the-right-person-at-the-right-time problem. It is a question of getting computers to have a model of what you're doing, what your goals are, and matching that to what help they might have to offer.

So instead of seeing those ads as the obnoxious things they now are, think of them as the forerunner of something exciting. Think of them as the equivalent of your friend who is wise and ready to help at any time—only right now you have a very dumb and very annoying friend. Soon you will have smarter friends, lots more of them, and machines that can pick the best advice from that being proffered. And, of course, they don't have to actually be your friends. They can be the best and the brightest, pre-recorded and found with no effort, just in time. We understand enough of the science to do this now. Maybe soon we'll get tired of ads and start working on important things in AI.

BIOLOGY VERSUS CHOICE

THALIA WHEATLEY

Associate professor of psychological and brain sciences, Dartmouth College

In neuroscience, few single discoveries have the ability to stay news for long. However, in the aggregate, all lead to the emergence of perhaps the greatest developing news story: the widespread understanding that human thought and behavior are the products of biological processes. There is no ghost in the machine. In the public sphere, this understanding is dawning.

Consider the recent sea change in public opinion on homosexuality—the growing consensus that sexual orientation is not a choice. This transformation suggests that the scale is tipping from ancient intuition to an appreciation of biology with its inherent constraints and promises.

Every year, neuroscience reveals the anatomical and functional brain differences associated with expressing a given trait or tendency, whether psychopathy, altruism, extroversion, or conscientiousness. Researchers electrically stimulating one brain area cause a patient to experience a strong surge of motivation. Zapping a different area causes another to become less self-aware. Disease can disorient a patient's moral compass or create illusions of agency. Environmental influences, from what we eat to whom we see, provide inputs that interact with and shape our neural activity—the activity that instantiates all our thoughts, feelings, and actions. Finding by finding, the ghost in the machine is being unmasked as a native biological system.

But it is one thing to convince people that sexual orientation is not a choice. It is quite another to convince people that the whole dichotomy of biology versus choice makes no sense. Who besides the unmasked biological system is doing the "choosing"? Choosing whether to take medication is as much a biological phenomenon as the disease to be medicated. Choice is simply a fanciful shorthand for biological processes we do not yet apprehend. When we have communicated *that*—when references to choice occupy the same rhetorical space as the four humors—we will be poised to realize public policy in harmony with a scientific understanding of the mind.

HOW TO BE BAD TOGETHER

GLORIA ORIGGI

Philosopher, CNRS, Paris; author, *Qu'est-ce que la confiance?*

Completely unexpected—and hence interesting—was my reaction to the scientific news in Simon Gächter and Benedikt
Herrmann's compelling paper "Reciprocity, culture and human
cooperation: previous insights and a new cross-cultural experiment," in the *Philosophical Transactions of the Royal Society.*

The authors addressed a classic question in social science—
the tragedy of the commons, or the conflict between individual
interest and collective interest in dealing with common resources. This is a well-established conundrum in contemporary
behavioral economics and evolutionary sociobiology, usually
solved by (now) classic experimental results about cooperation,
trust, and altruistic punishment. A vast literature demonstrates
that direct and indirect reciprocity are important tools for ensuring human cooperation. People use "altruistic punishment"
to sustain cooperation—that is, they are willing to pay without
receiving anything back in order to sanction those who don't
cooperate and hence promote prosocial behavior. Yet Gächter
and Herrmann showed in their surprising paper that in some
cultures when people were tested in cooperative games (such as
the "public good game"), those who cooperated were punished,
not the free-riders.

In some societies, people prefer to act antisocially, and they
take actions to make sure others do the same! This means that
cooperation in societies is not always for the good: You can find

cartels of antisocial people who don't care at all for the common good and prefer to cooperate in keeping a status quo that suits them, even if the collective outcome is mediocre.

As an Italian with first-hand experience of living in a country where, if you behave well, you are socially and legally sanctioned, this news was exciting, even inspiring. Perhaps cooperation is not an inherent virtue of the human species. Perhaps, in many circumstances, we prefer to side with those who share our selfishness and weaknesses and to avoid prosocial, altruistic individuals. Perhaps it's not abnormal to live outside a circle of empathy. Perhaps cooperation for the collective worse is as widespread as cooperation for a better society!

PSYCHOLOGY'S CRISIS

ELLEN WINNER
Professor of psychology, Boston College

The field of psychology is experiencing a crisis. Our studies do not replicate. When *Science* recently published the results of attempts to replicate 100 studies, those results were not confidence-inspiring, to say the least.[*] The average-effect sizes declined substantially, and while 97 percent of the original papers reported significant *p* values, only 36 percent of the replications did.

The same difficulty in reproducing findings is found in other scientific fields. Psychology is not alone. We know why so many studies that don't replicate were published in the first place: because of the intense pressure to publish in order to get tenure and grants and teach fewer courses—and because of journals' preference for publishing counterintuitive findings over less surprising ones. But it is worth noting that one-shot priming studies are far more likely to be flukes than longitudinal descriptive studies (e.g., studies examining changes in language in the second year of life) and qualitative studies (e.g., studies in which people are asked to reflect on and explain their responses and those of others).

In reaction to these jarring findings, journals are now changing their policies. No longer will they accept single studies with

---

[*] "Estimating the reproducibility of psychological science," 349: 6251, 28 Aug. 2015.

small sample sizes and p values hovering just below .05. But this is only the first step. Because new policies will result in fewer publications per researcher, universities will have to change their hiring, tenure, and reward systems, and so will granting and award-giving agencies. We need to stop the lazy practice of counting publications and citations, and instead read critically for quality. That takes time.

Good will come of this. Psychology will report findings that are more likely to be true, less likely to lead to urban myths. This will enhance the field's reputation and, more important, our understanding of human nature.

THE TRUTHINESS OF SCIENTIFIC RESEARCH

JUDITH RICH HARRIS

Independent investigator and theoretician; author, *The Nurture Assumption*

The topic itself is not new. For decades, there have been rumors about famous historical scientists like Newton, Kepler, and Mendel; the charge was that their research results were too good to be true. They must have faked the data, or at least prettied it up a bit. But Newton, Kepler, and Mendel nonetheless retained their seats in the Science Hall of Fame. The usual reaction of those who heard the rumors was a shrug. So what? They were right, weren't they?

What's new is that nowadays everyone seems to be doing it, and they're not always right. In fact, according to John Ioannidis, they're not even right most of the time. John Ioannidis is the author of a paper titled "Why Most Published Research Findings Are False," which appeared in a medical journal in 2005.* Nowadays this paper is described as "seminal," but at first it received little attention outside the field of medicine, and even medical researchers didn't seem to be losing sleep over it.

Then people in my own field, psychology, began to voice similar doubts. In 2011, the journal *Psychological Science* published a paper titled "False-positive Psychology: Undisclosed Flexibility in Data Collection and Analysis Allows Presenting Anything

---

* http://journals.plos.org/plosmedicine/article?id=10.1371/journal.pmed.0020124

as Significant." In 2012, the same journal published a paper on "the prevalence of questionable research practices."[†] In an anonymous survey of more than 2,000 psychologists, 53 percent admitted that they had failed to report all of a study's dependent measures, 38 percent had decided to exclude data after calculating the effect it would have on the outcome, and 16 percent had stopped collecting data earlier than planned because they had gotten the results they were looking for.

The final punch landed in August 2015. The news was published first in the journal *Science* and quickly announced to the world by the *New York Times*, under a headline that was surely facetious: "PSYCHOLOGISTS WELCOME ANALYSIS CASTING DOUBT ON THEIR WORK." The article itself painted a more realistic picture. "The field of psychology sustained a damaging blow," it began. "A new analysis found that only 36 percent of findings from almost 100 studies in the top three psychology journals held up when the original experiments were rigorously redone." On average, effects found in the replications were only half the magnitude of those reported in the original publications.

Why have things gone so badly awry in psychological and medical research? And what can be done to put them right again?

I think there are two reasons for the decline of truth and the rise of truthiness in scientific research. First, research is no longer something people do for fun, because they're curious. It has become something that people are required to do, if they want a career in the academic world. Whether they enjoy it or not, whether they are good at it or not, they've got to turn out papers every few months or their career is down the tubes. The rewards for publish-

---

† https://www.cmu.edu/dietrich/sds/docs/loewenstein/Mea-sPrevalQuestTruthTelling.pdf

ing have become too great relative to the rewards for doing other things, such as teaching. People are doing research for the wrong reasons: not to satisfy their curiosity but to satisfy their ambitions. There are too many journals publishing too many papers. Most of what's in them is useless, boring, or wrong. The solution is to stop rewarding people on the basis of how much they publish. Surely the tenure committees at great universities could come up with other criteria on which to base their decisions!

The second thing that has gone awry is the vetting of research papers. Most journals send out submitted manuscripts for review. The reviewers are unpaid experts in the same field, who are expected to read the manuscript carefully, make judgments about the importance of the results and the validity of the procedures, and put aside any thoughts of how the publication of this paper might affect their own prospects. It's a hard job that has gotten harder over the years, as research has become more specialized and data analysis more complex. I propose that this job be performed by paid experts—accredited specialists in the analysis of research. Perhaps this could provide an alternative path into academia for people who don't particularly enjoy the nitty-gritty of doing research but love ferreting out the flaws in the research of others.

In Woody Allen's movie *Sleeper*, set 200 years in the future, a scientist explains that people used to think that wheat germ was healthy and that steak, cream pie, and hot fudge were unhealthy—"precisely the opposite of what we now know to be true." It's a joke that hits too close to home. Bad science gives science a bad name. Whether wheat germ is or isn't good for people is a minor matter. But whether people believe in scientific research or scoff at it is of crucial importance to the future of our planet and its inhabitants.

BLINDED BY DATA

GARY KLEIN

Psychologist; senior scientist, MacroCognition LLC; author, *Seeing What Others Don't*

The 23 October 2015 issue of the journal *Science* reported a feel-good story about how some children in India had received cataract surgery and were able to see. On the surface, nothing in this incident should surprise us. Ready access to cataract surgery is something we take for granted. But the story is not that simple.

The children had been born with cataracts. They had never been able to see. By the time their condition was diagnosed—they came from impoverished and uneducated families in remote regions—the regional physicians had told the parents that it was too late for a cure because the children were past a critical period for gaining vision. Nevertheless, a team of eye specialists arranged for the cataract surgery to be performed even on teenagers. Now, hundreds of formerly blind children are able to see. After having the surgery four years earlier, one young man of twenty-two can ride a bicycle through a crowded market.

The concept of a critical period for developing vision was based on studies that David Hubel and Torsten Wiesel performed on cats and monkeys. The results showed that without visual signals during a critical period of development, vision is impaired for life. For humans, this critical window was thought to close tight by the time a child was eight years old. (For ethical reasons, no comparable studies were run on humans.) Hubel and Wiesel were awarded the Nobel Prize in 1981 for their work.

And physicians around the world stopped performing cataract surgery on children older than eight. The data were clear. But they were wrong. The results of the cataract surgeries on Indian teenagers disprove the critical-period data.

In this light, an apparent "feel-good" story becomes a "feel-bad" story about innumerable other children who were denied cataract surgery because they were too old. Consider all the children who endured a lifetime of blindness because of excessive faith in misleading data.

The theme of excessive faith in data was illustrated by another 2015 news item. Brian Nosek and a team of researchers set out to replicate 100 high-profile psychology experiments that had been performed in 2008. They reported their findings in the 28 August 2015 issue of *Science*. Only about a third of the original findings were replicated and even for these, the effect size was much smaller than the initial report.

Other fields have run into the same problem. A few years ago the journal *Nature* reported that most of the cancer studies selected for review could not be replicated. In October 2015, *Nature* devoted a special issue to exploring various ideas for reducing the number of non-reproducible findings. Many others have begun examining how to reduce the chances of unreliable data.

I think this is the wrong approach. It exemplifies the bedrock bias: a desire for a firm piece of evidence that can be used as a foundation for deriving inferences.

Scientists appreciate the tradeoff between Type I errors (detecting effects that aren't actually present—false positives) and Type II errors (failing to detect an effect that is present—false negatives). When you put more energy into reducing Type I errors, you run the risk of increasing Type II errors, missing findings and discoveries. Thus we might change the required significance level from .05 to .01, or even .001, to reduce the chances of a false positive, but in so doing we would greatly increase the false negatives.

The bedrock bias encourages us to make extreme efforts to eliminate false positives, but that approach would slow progress. A better perspective is to give up the quest for certainty and accept the possibility that any datum may be wrong. After all, skepticism is a mainstay of the scientific enterprise.

I recall a conversation with a decision researcher who insisted that we cannot trust our intuitions; instead, we should trust the data. I agreed that we should never trust intuitions (we should listen to our intuitions but evaluate them), but I didn't agree that we should trust the data. There are too many examples, as described above, where the data can blind us.

What we need is the ability to draw on relevant data without committing ourselves to their validity. We need to be able to derive inferences, make speculations, and form anticipations in the face of ambiguity and uncertainty. And to do that, we will need to overcome the bedrock bias—to free ourselves from the expectation that we can trust the data.

I'm not arguing that it's OK to get the research wrong—witness the consequence of all the Indian children who suffered unnecessary blindness. My argument is that we shouldn't ignore the possibility that the data might be wrong. The team of Indian eye specialists responded to anecdotes about cases of recovered vision and explored the possible benefits of cataract surgery past the critical period.

The heuristics-and-biases community has done an impressive job of sensitizing us to the limits of our heuristics and intuitions. Perhaps we need a parallel effort to sensitize us to the limits of the data—a research agenda demonstrating the kinds of traps we fall into when we trust the data too much. This agenda might examine the underlying causes of the bedrock bias, and possible methods for de-biasing ourselves. A few cognitive scientists have performed experiments on the difficulty of working with ambiguous data, but I think we need more: a larger, coordinated research program.

Such an enterprise would have implications beyond the scientific community. We live in an era of Big Data, an era in which quants are taking over Wall Street, an era of evidence-based strategies. In a world that is increasingly data-centered, there may be value in learning how to work with imperfect data.

THE EPISTEMIC TRAINWRECK OF SOFT-SIDE PSYCHOLOGY

PHILIP TETLOCK

Psychologist and political scientist; Annenberg University
Professor, University of Pennsylvania; co-author (with Dan Gardner),
Superforecasting

Thirty-five years ago, I was an insecure assistant professor at the University of California at Berkeley, and a curmudgeonly senior colleague from the hard-science side of psychology took me aside to warn me that I was wasting whatever scientific talent I might have. My field, broad-brushed as the soft side of psychology, was well intentioned but premature. Soft-siders wanted to help people but they hadn't a clue how to do it.

Now I get to play curmudgeon. The recent wave of disclosures about the non-replicability of many soft-side research phenomena suggests that my skeptical elder knew more than I then realized. The big soft-side scientific news is that a disconcertingly large fraction of it does not bear close scrutiny. The exact fraction is hard to gauge; my current guess is at least 25 percent and perhaps as high as 50. But historians of science will not have a hard time portraying this epistemic trainwreck as retrospectively inevitable. Social psychology and overlapping disciplines evolved into fields that incentivized scholars to get over the talismanic $p < .05$ significance line to support claims to original discoveries, and disincentivized the grunt work of assessing replicability and scoping out boundary conditions. And as Duarte *et al.* point out in *Behavioral and Brain Sciences* ("Political Diversity Will Improve Social Psychological Science"), the growing political homogeneity of the field selectively incentivized the production of counterintuitive findings that would jar the public into realizing how deeply unconsciously unfair the social order is. This has proved a dangerous combination.

In our rushed quest to establish our capacity for surprising smart outsiders while also helping those who had long gotten the short end of the status stick, soft-siders forgot the normative formula that Robert Merton offered in 1942 for successful social science: the CUDOS (Communalism, Universalism, Disinterestedness, Originality, and Skepticism) norms for protecting us from absurdities like Stalinist genetics and Aryan physics. The road to scientific hell is paved with political intentions, some well intentioned, some maniacally evil. If you value science as a purely epistemic game, the effects are equally corrosive. When you replace the pursuit of truth with the protection of dogma, you get politically/religiously tainted knowledge. Mertonian science imposes monastic discipline; it bars even flirting with ideologues.

I timed my birth badly, but those entering the field today should see the trainwreck as a gold mine. My generation's errors are their opportunities. Silicon-Valley-powered soft science gives us the means of enforcing Mertonian norms of radical transparency in data collection, sharing, and interpretation. We can now run forecasting tournaments around Open Science Collaborations in which clashing schools of thought ante up their predictions on the outcomes of well-designed, large-sample-size studies, earning or losing credibility as a function of rigorously documented track records rather than who can sneak what by which sympathetic editors. Once incentives and norms are reasonably aligned, soft science should firm up fast. Too bad I cannot bring myself to believe in reincarnation.

SCIENCE ITSELF

PAUL BLOOM

Ragen Professor of Psychology and Cognitive Science, Yale University; author, *Just Babies: The Origins of Good and Evil*

The most exciting recent scientific news is about science itself: how it is funded, how scientists communicate with another, how findings get distributed to the public—and how it can go wrong. My own field of psychology has been Patient Zero here, with well-publicized cases of fraud, failures to replicate important studies, and a host of concerns, some of them well founded, about how we do our experiments and analyze our results.

There's a lot to complain about with regard to how this story has played out in the popular press and over social media. Psychology—and particularly social psychology—has been unfairly singled out. The situation is at least as bad in other fields, such as cancer research. More important, legitimate concerns have been exaggerated and used by partisans on both the left and the right to dismiss any findings that don't fit their interests and ideologies.

But it's a significant story, and a lot of good can come from it. It's important for non-scientists to have some degree of scientific literacy, and this means more than a familiarity with certain theories and discoveries. It requires an appreciation of how science works, and how it stands apart from other human activities, most notably religion. A serious public discussion of what scientists are doing wrong and how they can do better will not only lead to better science but will help advance scientific understanding more generally.

A COMPELLING EXPLANATION FOR SCIENTIFIC MISCONDUCT

LEO M. CHALUPA

Neurobiologist; Vice President for Research, George Washington University

There were plenty of remarkable discoveries this past year, my favorite demonstrating that running promotes the generation of new neurons in the aging brain. That prompted me to get back on the treadmill. But the big scientific news story for me was not any single event; rather, I have been struck by the emergence over the past several years of two related trends in the scientific world. Neither of these has made front-page news, although both are well known to those of us in the science business.

The first of these is the apparent increase in the reported incidence of research findings that cannot be replicated. The causes for this are myriad. In some cases, it's simply because some vital piece of information, required to repeat a given experiment, has been inadvertently (or at times intentionally) omitted. More often, it is the result of sloppy work—poor experimental design, inappropriate statistical analysis, lack of appropriate controls.

But there is also evidence that scientific fraud is on the increase. A number of sensational cases came to light in 2015, and the incidence of retracted, withdrawn, and corrected scientific papers has increased steadily over the past decade. Indeed, some pharmaceutical companies have decided to no longer rely on the results of published studies, fearing that many of these are not trustworthy. Some have argued that the growing scientific misconduct reflects the use of new technologies for uncovering it—such as programs that check for plagiarism. Technological advances have certainly played a role, but in my opinion this doesn't entirely explain the deluge of failures to replicate.

A more compelling explanation is the fiercely competitive nature of science, which has accelerated in recent years. Grants are much harder to get funded, so that even applications ranked by peer review as "very good" are no longer above the pay line. So faculty at research universities must spend the bulk of their time writing grant proposals, with the probability of funding often being less than 10 percent. At the same time, the most highly ranked scientific journals have increased their rejection rates considerably; acceptance rates in the neighborhood of 5 percent are not uncommon. Moreover, the editors of the top journals (based on the recommendations of anonymous referees) often request additional experiments, which requires considerably more time, expense, and effort with no guarantee that the final product will be accepted for publication. For practitioners of science the level of stress has never been greater. Is it any wonder that some succumb to a shortcut method to success by "fudging" their results to get a competitive edge?

I'm not suggesting that simply increasing the amount of available funding for research would abate this problem. The solution will require a multifaceted approach. One step that should be seriously considered is a substantial increase in the penalty for scientific fraud, with jail time for those who have wasted precious research dollars. Today, those who are caught often get away relatively unscathed, in some instances receiving a substantial buyout package to leave their place of employment, because of the nature of the university tenure system. This trend must be reversed to ensure the viability of the scientific enterprise.

SUB-PRIME SCIENCE

NICHOLAS HUMPHREY

Emeritus Professor of Psychology, London School of Economics;
Visiting Professor of Philosophy, New College of the Humanities;
Senior Member, Darwin College, Cambridge; author, *Soul Dust*

In August 2015, Brian Nosek and the Open Science Collaboration published a report in *Science* on the replicability of findings previously published in top-rank psychology journals: "We conducted replications of 100 experimental and correlational studies . . . using high-powered designs and original materials when available." Only 36 percent of the replications were successful. Among the findings that didn't replicate were these:

- "People are more likely to cheat after they read a passage informing them that their actions are determined and thus that they don't have free will."
- "People make less severe moral judgments when they've just washed their hands."
- "Partnered women are more attracted to single men when they're ovulating."

These particular findings may not be game-changing. But they have been widely cited by other researchers (including me).

In many cases there may well be innocent explanations for why the original study gave the unreliable results it did. But in more than a few cases it can only be put down to slipshod research, too great haste to publish, or outright fraud. Worryingly, the more newsworthy the original finding, the more likely it could not be replicated. Insiders have likened the situation to a trainwreck.

John Brockman likes to quote Stewart Brand: "Science is the only news. When you scan through a newspaper or magazine, all the human interest stuff is the same old he-said-she-said, . . . a pathetic illusion of newness. . . . Human nature doesn't change much; science does." But we have here a timely reminder that the distinction between science and journalism is not—and has never been—as clear-cut as Brand imagines.

The reality is that science itself has always been affected by "the human interest stuff." Personal vendettas, political and religious biases, stubborn adherence to pet ideas have in the past led even some of the greatest scientists to massage experimental data and skew theoretical interpretations. Happily, the body of scientific knowledge has continued to live and grow despite such human aberrations. In general, scientists continue to play by the rules.

But we must not be complacent. The professional culture is changing. In many fields, and not of course only in psychology, science is becoming more of a career path than a noble vocation, more of a feeding trough than a chapel of truth. Sub-prime journals are flourishing. Bonuses are growing. After the disgrace of the bankers, science must not be next.

THE INFANCY OF META-SCIENCE

JONATHAN SCHOOLER
Professor, Department of Psychological and Brain Sciences,
UC Santa Barbara

A defining feature of science is its ability to evolve in response to new developments. Historically, changes in technological capacities, quantitative procedures, and scientific understanding have all contributed to large-scale revisions in the conduct of scientific investigations. Pressure is mounting for further improvements. In disciplines such as medicine, psychology, genetics, and biology, researchers have been confronting findings that are not as robust as they initially appeared. Such shrinking effects raise questions not only about the specific findings they challenge but more generally, about the confidence we can have in published results that have yet to be re-evaluated.

In attempting to understand its own limitations, science is fueling the consolidation of an emerging new discipline: meta-science. Meta-science, the science of science, attempts to use quantifiable scientific methodologies to show how current scientific practices influence the truth of scientific conclusions. This endeavor is joining the agendas of a variety of fields, including medicine, biology, and psychology—each seeking to understand why some initial findings fail to fully replicate. Meta-science has its roots in the philosophy of science and the study of scientific methods but is distinguished from the former by its reliance on quantitative analysis and from the latter by its broad focus on the general factors contributing to the limitations and successes of scientific investigations.

An ambitious meta-scientific study was recently published in *Science* by Brian Nosek and the Open Science Collaboration. A large-scale effort in psychology sought to replicate 100 "quasi-randomly" selected studies from three premier journals and found that less than half reached traditional levels of significance when replicated. This study is noteworthy because it directed the lens of science not at any particular phenomenon but rather at the process of science itself. In this sense, it represents one of the first major implementations of evidence-based meta-science.

Although I'm enthusiastic about the meta-scientific goals this study exemplifies, I worry that major limitations in its design and implementation may have produced a misleadingly pessimistic assessment of the health of the field of psychology. Numerous factors may have contributed to an underestimation of the reliability of the findings, including variations in the skills and motivations of the replicating scientists, limitations in the statistical power of the replications, and perhaps most important, questions regarding the fidelity with which the original methods were reproduced. Although the authors attempted to vet their replication procedure with the originating lab, many of the replicated studies were conducted without the originating lab's endorsement, and these unapproved efforts disproportionately contributed to the low replication estimate.

Even the studies that used procedures approved by the originating laboratories may have been lacking in fidelity. For example, one of the more well-known findings that failed to replicate involved the observation that exposing people to an anti–free-will message can increase cheating. I'm particularly familiar with this example (and perhaps biased to defend it), as I was a co-author of the original study. Although we signed off on the replication protocol, we subsequently discovered a small but important detail that was left out of the replicating procedure. In the original study, but not the replication, the anti–free-will message was framed as part of an entirely different study. We have recently found that people are less likely to change their beliefs about free will when the anti–free-will message is introduced as part of the same study. Apparently people are reluctant to change their mind on this important topic if they feel coerced to do so. In this context, it is notable that in the replication study, the anti–free-will message failed to significantly discourage participants from believing in free will in the first place, and thus could hardly have been expected to produce the further ramification of increased cheating. I suspect that a big portion of failures to replicate may involve the omission of similar small but important methodological details.

As the emerging field of meta-science moves forward, it will be important to refine techniques for understanding how disparities between original studies and replications may contribute to difficulties in reproducing results. Increasing the transparency of originally conducted studies, through methods such as detailed pre-registration, is likely to make it easier for replication teams to understand precisely how the project was originally implemented. However, it will also be important to develop methods for evaluating the fidelity of the reproductions themselves.

Another important next step for meta-science is the implementation of prospective replication experiments that systematically investigate how new hypotheses fare when tested repeatedly across laboratories. Prospective replication experiments will help to overcome potential biases inherent in selecting which published studies to replicate, while simultaneously illuminating various factors that may govern the replicability of scientific findings, including variations in population sample, researcher investment, and reproduction fidelity.

As we adopt a more meta-scientific perspective, researchers will increasingly appreciate that just as a single study cannot irrefutably demonstrate the existence of a phenomenon, neither can a single failure to replicate disprove it. Over time, scientists will likely become more comfortable with meticulously documenting and (ideally) pre-registering all aspects of their research. They will see the replication of their work not as a threat to their integrity but as testament to their work's importance. They will recognize that replicating other findings is an important component of their scientific responsibilities. They will refine replication procedures not only to discern the robustness of findings but also to understand their boundary conditions and the reasons they sometimes (often?) decline in magnitude. Even if history shows that the original foray into meta-science was significantly lacking, ultimately meta-science will surely offer deep insights into the nature of the scientific method itself.

THE DISILLUSION AND THE DISAFFECTION OF POOR WHITE AMERICANS

RICHARD NISBETT

Theodore M. Newcomb Distinguished Professor of social psychology; codirector, Culture and Cognition program, University of Michigan at Ann Arbor; author, *Mindware*

> The mortality rate for whites 45 to 54 years old with no more than a high school education increased by 134 deaths per 100,000 people from 1999 to 2014.
>
> —*NEW YORK TIMES*, NOV. 2, 2015

Over the past fifteen years or so, the mortality rate for poorly educated middle-aged whites living in the U.S. South and West increased significantly. Mortality did not increase for middle-aged blacks, Hispanics, or any other ethnic group, nor for whites in other regions of the country, nor for poorly educated whites in other rich countries. The death rates that are most elevated are those for suicide, cirrhosis of the liver, heroin overdose, and other causes suggesting self-destructive behavior.

There is some controversy about just how great the increase in mortality is, and whether it holds only for women or for both men and women, but there is no debate about the fact that late-middle-aged poor American whites are doing relatively badly with respect to mortality rates—both as compared with other Americans and as compared with people in rich countries generally. And the warning signs that something is very wrong with white people at the bottom of the American economic ladder are mounting rapidly.

The worsening plight of poor white Americans highlighted by the *Times* article on the mortality findings by Princeton economists Angus Deaton and Anne Case is by no means limited to just the South and West. Researchers from political science to neuroscience have been uncovering ever more disturbing facts about whites at the bottom of the U.S. socioeconomic ladder. Charles Murray, in his book *Coming Apart* (2012), showed that between 1960 and 2010 the bottom 30 percent of Americans in terms of socioeconomic status (SES) experienced a collapse in social capital. The rate of children from broken marriages and living with a single parent increased tenfold over that period—to 25 percent. The rate of children living with both biological parents when the mother was forty years old plummeted from 95 percent to 30 percent. The fraction of people having no involvement in any secular or religious organization more than doubled—to 34 percent. The percentage of prime-working-age males not in the workforce increased threefold, to 12 percent. The percentage of men not making enough to support a household of two more than doubled, to 30 percent. The percentage of males in state and federal prisons grew almost fivefold.

Murray examined the same variables for the 20 percent of the white population with the highest socioeconomic status. For none of these variables was there a notable worsening over the fifty-year period.

Sociologist Sean Reardon examined the gap between the academic achievement of the top 10 percent of the SES spectrum and the bottom 10 percent between the late 1940s and the early 2010s. He also examined the black/white gap in academic achievement over that time span. At the beginning of the period, the black/white gap was double the SES gap. At the end of that period, the SES gap was double the black/white gap. This crossover was due roughly in equal proportion to the gains of black children and the losses of lower-SES children.

Murray's claim that the welfare state is responsible for the lassitude and misery of the American lower class would appear to be ruled out by the fact that the social safety net is much stronger in Europe, and nothing there is close to the dire straits of those at the bottom of American society. It's easier to argue that it's the lack of a European-style safety net that has contributed to the American debacle.

So, what is responsible for the malaise at the bottom? Scientists have contributed little but speculation to this question. But a case could be made that one cause is that faith in the American dream, while still alive at the top of the economic pyramid, is disappearing at the bottom, and that this is true for primarily economic reasons. When I moved to Ann Arbor decades ago, a high-school educated worker on the line at Ford made enough money to support a family of four, own a three-bedroom home in the suburbs, have two cars and a boat, and buy a cottage in Northern Michigan. That's a higher standard of living for the poorly educated than was true in Europe then or now—or in the U.S. today. The poorly educated man today can expect to be an assistant manager of a chain store, a security guard, or a jack of all trades—occupations that barely support a single individual in modest fashion, let alone a family of four in comfort.

The disillusionment hypothesis explains why the support for Donald Trump's candidacy is greatest among ill-educated whites in the poorer, less cosmopolitan regions of the country. Trump's bombast, braggadocio, xenophobia, aggressiveness, and willingness to tell baldface lies is unnerving to anyone having a nodding acquaintance with the circumstances of the rise of fascism. Both Italian fascism and German Nazism achieved their greatest initial successes with the proletariat. In the case of Nazism, the greatest early gains were made among rural Protestant peasants.

Scientists have yet to develop convincing theories about what might alleviate the plight of poor whites at the bottom of the social ladder. Meanwhile we can only hope that the economic doldrums don't worsen, producing receptivity higher up the economic ladder to demagogues.

INEQUALITY OF WEALTH AND INCOME: A RUNAWAY PROCESS

S. ABBAS RAZA
Founding Editor, 3QuarksDaily.com

One of the biggest challenges facing us is the increasing disparity in wealth and income which has become obvious in American society in the last four decades or so, with all its pernicious effects on societal health. Thomas Piketty's data-backed *tour de force, Capital in the Twenty-First Century* (2013), gave us two alarming pieces of news about this trend: (1) Inequality is worse than we thought, and (2) it will continue to worsen because of structural reasons inherent in our form of capitalism, unless we do something.

The top 0.1 percent of families in America went from having 7 percent of national wealth in the late 1970s to having about 25 percent now. Over the same period, the income share of the top 1 percent of families has gone from less than 10 percent to more than 20 percent. And lest we think that even if wealth and income are more concentrated America is still the land of opportunity and those born with very little have a good chance to move up in economic class, a depressing number of studies show that according to standard measures of intergenerational mobility, the United States ranks among the least economically mobile of the developed nations.

Piketty shows that an internal feature of capitalism increases inequality: As long as the rate of return on capital (r) is greater than the rate of economic growth (g), wealth will tend to concentrate in a minority, and that the inequality $r > g$ always holds in the long term. And he is not some lone-wolf academic with an eccentric theory of inequality. Scores of well-respected economists have given ringing endorsements to his book's central thesis, including economics Nobel laureates Robert Solow, Joseph Stiglitz, and Paul Krugman. Krugman has written that

> Piketty doesn't just offer invaluable documentation of what is happening, with unmatched historical depth. He also offers what amounts to a unified field theory of inequality, one that integrates economic growth, the distribution of income between capital and labor, and the distribution of wealth and income among individuals into a single frame.[*]

The only solution to this growing problem, it seems, is the redistribution of the wealth concentrating within a tiny elite using instruments like aggressive progressive taxation (such as exists in some European countries that show a much better distribution of wealth), but the difficulty here is the obvious one that political policymaking is itself greatly affected by the level of inequality. This vicious positive-feedback loop makes things even worse. It is clearly the case now in the United States that not only can the rich hugely influence government policy directly but also that elite forces shape public opinion and affect election outcomes with large-scale propaganda efforts through media they own or control. This double-edged sword attacks and shreds democracy itself.

The resultant political dysfunction makes it difficult to address our most pressing problems—for example, lack of oppor-

---

[*] "Why We're in a New Gilded Age," NY Review of Books, May 8, 2014.

tunity in education, lack of availability of quality healthcare, man-made climate change, and not least the indecent injustice of inequality itself. I'm not sure if there is any way to stop the growth in inequality we have seen in the last four or five decades anytime soon, but I do believe it is one of the important things we have learned more about in the last couple of years. Unfortunately the news is not good.

THE AGE OF VISIBLE THOUGHT

PETER GABRIEL
Singer-songwriter, musician, humanitarian activist

It now seems inevitable that the decreasing cost and increasing resolution of brain-scanning systems, accompanied by a relentless increase in computer power, will take us soon to the point where our own thinking may be visible, downloadable, and open to the world in new ways.

It was the news that brain scanners are starting to be developed at consumer price levels that obsessed me this year.

Through the work of Mary Lou Jepsen, I was introduced to the potential of brain-reading devices and learned that patterns generated while watching a succession of varied videos would provide the fundamental elements to connect thought to image. A starting point was the work pioneered at Jack Gallant's Lab at UC Berkeley in 2011, which proved that the patterns of brain activity from fMRI scanners when a subject was viewing an assortment of videos would enable thoughts to be translated into digital images.

Recording more and more images and corresponding brain patterns boosts the vocabulary in the individual's visual dictionary of thought. Accuracy greatly increases with the quantity and quality of data and of the decoding algorithms. Jepsen persuaded me that this is realizable within a decade, within the cost range of consumer electronics, and in a form that appeals to non-techies. Laborious techniques and huge, power-hungry, multimillion-dollar systems based on magnetic fields will be succeeded by optical techniques where the advantages of consumer electronics can assert themselves; the power of AI algorithms will do the rest. This science-fiction future is not only realizable but, because of enormous potential benefits, will inevitably be realized.

And so here we are: Our thoughts themselves are about to take a leap out of our heads, from our brains to computer, to the Internet, and to the world. We are entering the Age of Visible (and Audible) Thought. This will surely affect human life as deeply as any technology our imagination has yet devised, or any evolutionary advance.

The essence of who we are is contained in our thoughts and memories, which are about to be opened like tin cans and poured onto a sleeping world. Inexpensive scanners would enable us to display our thoughts and access those of others who do the same. The consequences and ethics of this have barely been considered. I imagine the pioneers of this research enjoying a heady Oppenheimer cocktail of anticipation and foreboding, of exhilaration and dread. Our task is to ensure that they do not feel alone or ignored.

One giant tech company is believed to have already backed off exploring the development of brain reading for Visual Thought, apparently fearing the potentially negative repercussions and controversy over privacy. The emergence of this suite of technologies will have enormous impact on the everyday ways we live and interact and can clearly transform, positively and negatively, our relationships, aspirations, work, creativity, and techniques for extracting information. Those not comfortable swimming in these transparent waters will not flourish. Perhaps we'll need to create "swimming lessons" to teach us how to be comfortable being open, honest, and exposed—ready to navigate these waters of visible thought.

What else happens in a World of Visible Thought? One major difference is that as thought becomes closer and closer to action, with shorter feedback loops accelerating change, timescales collapse and the cozy security blanket of a familiar slowness evaporates. A journey for my grandfather from London to New York shrank from a perilous three weeks to a luxurious three hours for my generation on the Concorde. Similarly, plugging thought directly into the material world will all but eliminate the comfort of time lag. If I look outside at the streets, the buildings, the cars, I am just looking at thought turned into matter, the idea in its material form. With 3D printing and robotics, that entire process can become nearly instantaneous.

The past year has witnessed robots building bridges and houses, but these currently work from 3D blueprints. Soon we'll be able to plug in the architect directly and, with a little bit of fine tuning, see her latest thoughts printed and assembled into a building immediately. The same goes for film, for music, and for every other creative process. Barriers between imagination and reality are about to burst open. Do we ignore it, or do we get into boat-building, like Noah? Here comes the flood. . . .

OUR CHANGING CONCEPTIONS
OF WHAT IT MEANS TO BE HUMAN

HOWARD GARDNER

Hobbs Professor of Cognition and Education, Harvard Graduate
School of Education; author, *Truth, Beauty, and Goodness Reframed*

We live at a time of great, perhaps unprecedented, advances in
digital technology (hardware/software) and biological (genetic/
brain) research and applications. It's easy to see these changes as
wholly or largely positive, although as a card-carrying member
of the pessimists' society I can easily point to problematic aspects
as well. But irrespective of how full (or empty) you believe the
glass to be, a powerful question emerges: To what extent will
our conceptions of what it means to be human change?

History records huge changes in our species over the last
5,000 years or so—and presumably prehistory would fill in
the picture. But scholars have generally held the view that the
fundamental nature of our species—the human genome, so to
speak—has remained largely the same for at least 10,000 years
and possibly much longer. As Marshall McLuhan argued, technology extends our senses, it does not fundamentally change
them. Once one begins to alter human DNA (for example,
through CRISPR) or the human nervous system (by inserting
mechanical or digital devices), we are challenging the very definition of what it means to be human. And once one cedes high-
level decisions to digital creations, or these artificially intelligent
entities cease to follow the instructions programmed into them

and rewrite their own processes, our species will no longer be dominant on this planet.

In a happy scenario, such changes will take place gradually, even imperceptibly, and they may lead to a more peaceful and even happier planet. But as I read the news of the day, and of the last quarter century, I discern little preparedness on the part of human beings to accept a lesser niche, let alone to follow Neanderthals into obscurity. And so I expect tomorrow's news to highlight human resistance to fundamental alterations in our makeup, and quite possibly feature open warfare between old and newly emerging creatures. But there will be one difference from times past: Rather than looking for insights in the writings of novelists like Aldous Huxley or George Orwell or Anthony Burgess, we'll be eavesdropping on the conversations among members of the third culture.

COMPLETE HEAD TRANSPLANTS

KAI KRAUSE
Software pioneer; philosopher; author, *A Realtime Literature Explorer*

Early this year an old friend, a professor of neurology, sent me an article from a medical journal, *Surgical Neurology International*—at first glance, predictably, a concoction of specialist language. The "Turin Advanced Neuromodulation Group" is describing "Cephalosomatic Anastomosis" (CSA), to be performed with "a nanoknife made of a thin layer of silicon nitride with a nanometer sharp-cutting edge."

Only slowly it becomes clear that they are talking about something rather unexpected: "Kephale," Greek for head; "Somatikos," Greek for body; "Anastasis," Latin for resurrection—that prosaic CSA stands for *a complete head transplantation*. And that reverberated with me, the implications being literally mind-boggling.

The thought of a functioning brain reconnected to an entirely new body opens up any number of speculations. And has done so in countless sci-fi books and B movies. But there is a lot to consider.

The author, Italian surgeon Sergio Canavero, announced a few months later that he had a suitable donor for the head part and suddenly made it sound quite real, adding tangible details: The operation, to be performed in 2017, would take place in China, require a team of 150 specialists, take 36+ hours, and cost \$15+ million. Then it hit the mass media. Many responses revolved around the ethics of such an action, using the F-word a lot (and I mean "Frankenstein") and debating the scientific details of the spinal-cord fusion.

My stance on the ethical side is biased by a personal moment: In the mid-nineties I visited Stephen Hawking in Cambridge for a project and he later visited in Santa Barbara—both interactions, up close, left me with an overpowering impression. There was that metaphor of "the mind trapped in a body," playing out in all its deep and poignant extreme—the most intelligent of minds weighed down so utterly by the near useless shell of a body. A deep sadness would overcome anyone witnessing it—far beyond the Hollywood movie adaption.

There's the rub, then: Who could possibly argue against this man's choosing to lengthen his lifetime and gain a functioning body should such an option exist? Could anyone deny him the right to try, if medicine were up to the task? (Hawking is not a candidate even in theory, his head being afflicted by the disease as well, but he does serve as a touching and tragic example of that ethical side.)

Another personal connection for me is this: Critics called it "playing God" (imagine!). Human hubris. Where would the donors come from? Is this medicine just for the rich? Now, consider: The first such operation leads to the recipient's death after eighteen days. It is repeated, and the subsequent 100 operations lead to nearly 90 percent of the patients not surviving past the two-year mark. No, that is not a prognostication for CSA; I am recalling events from nearly fifty years ago. In December 1967, Christiaan Neethling Barnard performed the first human heart transplantation, the eyes of the world upon him; his face was subsequently plastered on magazine covers across the globe. I was ten and remembered his double-voweled name as much as the unfathomable operation itself.

He was met with exactly the same criticism, the identical ethical arguments.

After the dismal survival rate, the initial enthusiasm turned around, and a year later those condemning the practice were gaining. Only after the introduction of ciclosporin to vastly improve the immune-rejection issues did the statistics turn in his favor; tens of thousands of such operations have since been performed. Every stage of progress has had critical voices loudly extrapolating curves into absurdity; back then, as now, there is doomsday talk of "entire prison populations harvested for donors," and such.

Sadly, watching videos of Canavero on the Web is rather cringeworthy. Slinging hyperbolae such as "the world will never be the same," naming his protocols *heaven* and *gemini*, squashing a banana representing a damaged spinal cord versus a neatly sliced one to illustrate his ostensibly easy plan. He repeatedly calls it "fusing spaghetti" and even assessed the chances for his Russian donor at "90 percent to walk again."

The Guardian notes that "he published a book, *Donne Scoperte*, or *Women Uncovered*, that outlined his tried-and-tested seduction techniques." It seems clear that there is little place for levity when he belittles the details and glosses over the reality: millions of quadriplegic victims closely eyeing the chances of truly re-fusing spinal cords.

The story here is not about one celebrity poseur. In my view, it cannot happen by 2017, by far. But 2027, '37, '47? Looking backward, you can see the increase in complexity that makes it almost inevitable to think this *will* be possible. And then the truly interesting questions come into play. If phantom limbs bring serious psychological issues, what would an entire phantom body conjure up? The self-image is such a subtle process—the complexity of signals, fluids, and messenger chemistry—how could it all possibly attain a state remotely stable, let alone "normal"?

Christiaan Barnard, asked why anyone would choose such a risky procedure, replied, "For a dying person, a transplant is not a difficult decision. If a lion chases you to a river filled with crocodiles, you will leap into the water convinced you have a chance to swim to the other side. But you would never accept such odds if there were no lion."

Me, I dread even the dentist's waiting room. But thirty years hence, maybe I, too, would opt for the crocodiles. If Hawking can survive longer, by all means he should. Some other characters I can think of, their best hope lies in acquiring a new head. Thus I am of two minds about complete head transplants.

THE EN-GENDERING OF GENIUS

REBECCA NEWBERGER GOLDSTEIN

Philosopher, novelist; Visiting Professor, NYU; author, *Plato at the Googleplex*

For most of its history our species has systematically squandered its human capital by spurning the creative potential of half its members. Higher education was withheld from women in just about every place on Earth until the 20th century, with the few who persevered before then considered "unsexed." It's only been in the last few decades that the gap has so significantly closed that, at least in the U.S., more bachelor's degrees have been earned by women than by men since 1982, and since 2010 women have earned the majority of doctoral degrees. This recent progress only underscores the past's wasteful neglect of human resources.

Still, the gender gap has stubbornly perpetuated itself in certain academic fields, usually identified as STEM—science, technology, engineering, and mathematics—and this is as true in Europe as in the U.S. A host of explanations have been posed as to the continued male dominance—some only in nervous, hushed voices—as well as recommendations for overcoming the gap. If the underrepresentation of women in STEM isn't the result of innate gender differences in interests and/or abilities (this last, of course, being the possibility that can only be whispered), then it's important for us to overcome it. We've got enormously difficult problems to solve, both theoretical and practical, and it's lunacy not to take advantage of all the willing and able minds that are out there.

Which is why I found a 2015 article published in *Science* by Andrei Cimpian and Sarah-Jane Leslie big news.* First of all, their data show that the lingering gender gap shouldn't be framed in terms of STEM versus non-STEM. There are STEM fields—for example, neuroscience and molecular biology—that have achieved 50-percent parity in the number of PhDs earned by men and women in the U.S. And there are non-STEM fields—for example, music theory and composition (15.8 percent) and philosophy (31.4 percent)—where the gender gap rivals such STEM fields as physics (18 percent), computer science (18.6 percent) and mathematics (28.6). So that's the first surprise that their research delivers: that it's not science *per se* that, for whatever reasons, produces stubborn gender disparity. And this finding in itself somewhat alters the relevance of the various hypotheses offered for the tenacity of the imbalance.

The hypothesis that Leslie and Cimpian tested is one I've rarely seen put on the table and surely not in a testable form. They call it the FAB hypothesis—for field-specific ability beliefs. It focuses on the belief as to whether success in a particular field requires pure innate brilliance, the kind of raw intellectual power that can't be taught and for which no amount of conscientious hard work is a substitute. One could call it the Good-Will-Hunting quotient, after the 1997 movie featuring Matt Damon as a janitor at MIT who now and then, in the dead of night, pauses to put down his mop in order to effortlessly solve the difficult problems left scribbled on a blackboard.

To test the FAB hypothesis, the researchers sent out queries to practitioners—professors, postdocs, and graduate students—in

---

* "Expectations of brilliance underlie gender distributions across academic disciplines," 347: 6219, January 16, 2015.

leading U.S. universities, probing the extent to which the belief in innate brilliance prevailed in the field. In some fields, success was viewed as more a function of motivation and practice, while in others the Good-Will-Hunting quotient was more highly rated.

And here's the second surprise: the strength of the FABs in a particular field predicts the percentage of women in that field more accurately than other leading hypotheses, including field-specific variation in work/life balance and reliance on skills for systematizing vs. empathizing. In other words, what Cimpian and Leslie found is that the more success within a field was seen as a function of sheer intellectual firepower, with words such as "gifted" and "genius" not uncommon, the fewer the women. The FAB hypothesis cut cleanly across the STEM/non-STEM divide.

Cimpian and Leslie are careful to stress that they don't interpret their findings as indicating that the FAB hypothesis provides the sole factor behind the lingering gender gap, but simply argue that it is operative. And in follow-up studies, they also discuss informal evidence that raises the plausibility of the FAB hypothesis, including the number of fictional male geniuses inhabiting popular culture—from Sherlock Holmes to Dr. House to Will Hunting—compared to the number of female geniuses. The stereotype of the genius is overwhelmingly male. And when, I might add, a female genius is the subject, her femaleness itself becomes the focus as much as, or even more than, her genius. If genius is an aberration, then female genius is viewed as significantly more aberrational, since it's seen as an aberration of femaleness itself. Given such stereotypes, is it unlikely that fields that highlight innate genius would show lagging female numbers?

The authors were exclusively concerned with academic fields. But there is another area of human creativity in which words like "gifted" and "genius" are not uncommon, and that is the arts—including literature. Here, too, cold, hard statistics tell a story of persistent gender imbalance. For despite the great number of contemporary women writers, data compiled by VIDA, a women's literary organization, reveal that the leading American and British literary magazines—the kind whose very attention is the criterion for distinguishing between the important figures and the others—focus their review coverage on books written by men and commission more men than women to write about them. Might it be that the FAB hypothesis explains this imbalance as well, highlighting Cimpian's and Leslie's findings that the problem is not, essentially, one of STEM vs. non-STEM, nor of mathematical vs. verbal skills?

I realize that discussing the FAB hypothesis will be seen as small stuff compared to such big news as, say, the ice caps melting at a faster rate than anticipated. And that is why, in responding to this year's *Edge* Question, I first began to write about the ice caps. But perhaps the insignificant measure we assign to the underestimation of the creative potential of more than half our population is itself a manifestation of the problem. And what could be a greater boon to humanity than increasing the, um, manpower of those making important contributions, not only to science but to our culture at large?

DIVERSITY IN SCIENCE

GINO SEGRE
Professor of physics, emeritus, University of Pennsylvania; author,
Ordinary Geniuses

In a recent U.S. Supreme Court hearing regarding affirmative action in higher education, a Justice posed the question, "What unique perspective does a minority student bring to a physics class?" If physics were the work of robots, the answer would be none. But physics, as all of science, is approached with the biases and perspectives that make us human. Both disciplined and spontaneous, science is an intuitive as well as a systematic undertaking.

It is the people who comprise physics graduate schools, institutes, and faculty—their interests, their backgrounds, their agendas—who drive the direction of physics research and scholarship. And it does not take much imagination to picture what different directions this research might take, were the pool of scientists not heterogeneous.

Science has become increasingly collaborative in a way that makes diversity a paramount necessity. Until recently it was the work of single individuals, primarily white males from Northern Europe. It was rare to find a published paper with two authors and a paper with more than three was essentially unheard of. A change began at the time of World War II, and it has grown since then.

Large science collaborations encompassing diversity of gender, race, and ethnicity have become a new norm. ATLAS, a group that has been a major contributor to CERN's discovery of the Higgs boson, consists of 3,000 physicists from 175 institutions in 38 countries, working harmoniously together. Even though a single large instrument, such as a particle accelerator or a large telescope, is not required in biology, we see parts of the field moving in the same direction, with the Human Genome or Human Microbiome project. A different kind of complexity, the assembling of disparate parts of the pattern, is needed there.

The news is that science's success in such endeavors is creating a recognized model for international collaboration, the response to climate change being the most conspicuous example.

The diversity in approach engendered by diversity in background has been a powerful combination in science. It is no secret that in a fairly recent past, if physics graduate school admission had been based on achievement tests, the entering class would have been composed almost entirely of students from mainland China. Most of these schools believed that such a homogeneous grouping would not have benefited either the students or the field. It would have reinforced conformity rather than encouraging the necessary originality and entrepreneurship.

Science's future, both in the classroom and in research, is tied to an increased achievement of diversity in gender, race, ethnicity, and class. If this goal is not met, science will suffer.

THE DEMOCRATIZATION OF SCIENCE

MICHAEL SHERMER

Publisher, *Skeptic* magazine; monthly columnist, *Scientific American*;
Presidential Fellow, Chapman University; author, *The Moral Arc*

The biggest news story over the past quarter century—one that will continue to underlie all the currents, gyres, and eddies of individual sciences going forward—is the democratization of scientific knowledge. The first wave of knowledge diffusion happened centuries ago with the printing press and mass-produced books. The second wave took off after World War II, with the spread of colleges and universities and the belief that a higher education was necessary to being a productive citizen and cultured person. The third wave began a quarter century ago with the Third Culture: "Those scientists and other thinkers in the empirical world who, through their work and expository writing, are taking the place of the traditional intellectual in rendering visible the deeper meanings of our lives, redefining who and what we are," in John Brockman's 1991 description.

A lot has happened in twenty-five years. While some Third Culture products remain topical (AI, human genetics, cyberspace) and others have faded (chaos, fractals, Gaia), the culture of science as a redefining force endures and expands into the nooks and crannies of society through ever growing avenues of communication, pulling everyone in to participate. A quarter century ago, the third culture penetrated the public primarily through books and television; today third culture apostles spread the gospel through ebooks and audio books, digital books, and virtual libraries, blogs and microblogs, podcasts and videocasts, file-sharing and video-sharing, social networks and forums, MOOCs and remote audio and video courses, virtual classrooms, and even virtual universities.

The news is not just the new technologies of knowledge, however, but the acceptance by society's power brokers that third-culture products are the drivers of all other cultural products—political, economic, social, and ideological—and the realization of citizens everywhere that they, too, can be influential agents by absorbing and even mastering scientific knowledge.

This democratization of science changes everything, because it means we have unleashed billions of minds to solve problems and create solutions. The triumphs of the physical and biological sciences in the 20th century are now being matched by those in the social and cognitive sciences, because, above all else, we have come to understand that human actions, more than physical or biological forces, will determine the future of our species.

NEWS ABOUT SCIENCE NEWS

SHEIZAF RAFAELI

Professor, director, Center for Internet Research, University of Haifa, Israel

They say that news serves as the first draft of history and that reportage is just literature in a hurry. Both history and litera-ture have more patience and perspective than the often urgent work of science. So, what's new and what is news in the special domain of science? For me, the important news in this area is about news itself and the relation between news and science. The most important news about science is how transparent it is becoming.

News is both socially constructed and the construction of the social. It is socially constructed in that it is contextual, sub-jective, and ephemeral. And it constructs the social in that the work of news is to tell us who and what we are. We've known the social-construction aspect of news about politics and power since Plato's cave. Now we are learning to recognize more of the interplay between science and social construction with the emergence of transparent, open science news. Through the news about science news, we are learning how much science is socially constructed and how to deal with this fact.

The most important changes with regard to news are them-selves social. News, including science news, is collected, collated, curated, and consumed by ever growing circles of stakeholders. During our lifetime, or even the past decade, the science/news/society axis has been redrawn entirely. Instead of being a trickle-down so-called broadcast experience, news is now a bottom-up phenomenon. Fewer "invisible colleges" and many more public arenas for science. News about discoveries, innovations, controversies, and evidence are increasingly grass-roots generated and ranked, and universally more accessible. Economics of tuition and budget play a role, as does the evolving perception of the structure of knowledge.

Quite a few factors have formed these developments. Literacy is up, censorship is down. Access is up. Uniformity and control over news sources are down, even though algorithmic news curation and ranking are up. Thus, through the news about science and the new avenues for such news, expectations for the democratization of science, its funding and fruits, are all up. In fact, this venue—the public and cross-disciplinary conversations here on *Edge*—is a pleasant and prime example.

While attempts to control or filter news, including the news about science, have not slowed, the ability of regimes and authorities to put a lid on public knowledge of events and discoveries is falling apart. Sharing, in all its online forms, is up. Science news is a major case in point. The boundaries between scientific publishing and news enterprises are eroding. In this open and transparent environment, anti-intellectual and nonscientific phenomena such as conspiracy theories are less likely to hold for long. Truth just might have a chance.

This is not necessarily all good news. We should not let our guard down; problems and challenges are at both the high and low end. More transparent and participatory science may mean too much populism. Critical thinking about the organs and channels of news dissemination should continue. At the other, "high" end, monopolies still loom, not the least of them in scientific publishing. Concentrated ownership of media outlets is

still a threat, and in some locations growing. Attempts to manipulate the reporting of news, scientific literature, and learning
curricula in the service of an ideology, or of the powers that be,
or of special interests have not gone away. The loss of some traditional venues for news, the erosion of business models for others,
alongside the problems experienced by some of the organs of
scientific dissemination, are a continuing cause for concern. But
this is a transitional period, and the transition is in the right direction.

Whether the first draft of history or just "literature in a hurry,"
the important news is more in the eye of the beholder than set
in stone. Thus, the most encouraging news about science is that
there are many more eyes beholding, ranking, participating, and
reacting to the news of science.

THE BROADENING SCOPE OF SCIENCE

TANIA LOMBROZO
Associate professor of psychology, UC Berkeley

Every time you learn something, your brain changes. Children with autism have brains that differ from those of children who do not. Different types of moral decisions are associated with different patterns of brain activity. And when it comes to spiritual and emotional experience, neural activity varies with the nature of the experience.

In most ways, this is news that shouldn't be news—at least not in this century, or probably the last. Given what we already know about the brain and its relationship to behavior and experience, these claims don't tell us anything new. How could it have been otherwise? Any difference in behavior or experience must be accompanied by some change in the infrastructure that implements it, and we already know to look to the brain.

So why do neuroscientific findings of this type still make the news?

In part, it's because the details might be genuinely newsworthy—perhaps the specific ways in which the brain changes during learning, for example, can tell us something important about how to improve education. But there are two other reasons why neuroscientific findings about the mind might make headlines, and they deserve careful scrutiny.

The first reason comes down to what psychologist Paul Bloom calls "intuitive dualism." Intuitive dualism is the belief that mind and body, and therefore mind and brain, are fundamentally different—so different that it's surprising to learn of the carefully orchestrated correspondence revealed by the "findings" summarized above. It's wrong to *equate* mind with brain (perhaps, to quote Marvin Minsky, "the mind is what the brain does"), but we ought to reject the Cartesian commitments that underlie intuitive dualism no matter how intuitive they feel.

The second reason is because neuroscientific findings about the mind reveal the broadening scope of science. As our abilities to measure, analyze, and theorize have improved, so has the scope of what we can address scientifically. That's not new—what *is* new is the territory that now falls within the scope of science, including the psychology of moral judgment, religious belief, creativity, and emotion. In short, the mind and human experience. We're finally making progress on topics that once seemed beyond our scientific grasp.

Of course, it doesn't follow that science can answer all our questions. There are many empirical questions about the mind for which we don't yet have answers, and some for which we may never have answers. There are also questions that aren't empirical at all. (Contra Sam Harris, I don't think science—on its own—will ever tell us how we ought to live or what we ought to believe.) But the mind and human experience are legitimate topics of scientific study, and they're areas in which we're making remarkable, if painstaking, progress. That's good news to me.

Q-BIO

NIGEL GOLDENFELD

Center for Advanced Study Professor in Physics, University of Illinois
at Urbana-Champaign; director, NASA Astrobiology Institute for
Universal Biology, UIUC

The past year was a great one for science news, with two
coming-of-age stories, decades in the making, that are going to
capture the headlines in the years to come. The first item was
not in a newspaper, but if it had been, the splash headline would
read something like this: "A MATHEMATICIAN WITH A MODEL
ORGANISM! REALLY?"

You know what a mathematician is, but what about the term
"model organism"? It refers to biological investigators' deep
study, manipulation, and control of a carefully chosen organism—
for example, the fruit fly, *Drosophila melanogaster*, probably the
most widely used laboratory organism for studying multicellular
eukaryotes because generations of researchers have discovered
ingenious ways to manipulate its genetics and watch the cells as
they grow. It's almost unthinkable to associate a mathematician
with a model organism. So what's the story? Here's a vignette
that makes the point:

A few weeks ago, I had lunch with a mathematician colleague of mine who is an expert on differential equations and
dynamical systems. She writes papers with titles like "Non-
holonomic constraints and their impact on discretizations
of Klein-Gordon lattice dynamical models." During lunch,
she told me that her favorite model organism was *Daphnia*.

Daphnia are millimeter-sized planktonic organisms that live in ponds and rivers. They are so transparent that you can easily see inside them and watch what happens when they eat or drink (alcohol, for example, increases their heart rate). My colleague is part of a community that has developed ways to use mathematics to study ecology. They study populations, infectious diseases, ecosystem stability, and the competition for resources. Their work makes real predictions and they co-author papers with card-carrying ecologists.

A mathematician with a model organism marks the coming-of-age of Q-Bio, short for "quantitative biology." Generations of biologists have entered that subject in part to escape the horrors of calculus and other advanced mathematics. Yesterday's biology was a descriptive science; today, biology is becoming a quantitative and predictive discipline. One remarkable instance of the passing of the baton is that a principal leader of the public domain Human Genome Project was Eric Lander, founding director of the MIT-Harvard Broad Institute and a pure mathematician by training.

Applied mathematicians and theoretical physicists are developing new sophisticated tools to deal with other, non-genomic challenges in quantifying biology. One of these challenges is that the number of individuals in a community may be large, but not as large as the number of molecules of gas in your lungs, for example. So the traditional tools of physics based on statistical modeling have to be upgraded to deal with the large fluctuations encountered—such as in the number of proteins in a cell or individuals in an ecosystem.

Another is that living systems need an energy source; they are inherently out of thermodynamic equilibrium and so cannot be described by the century-old tools of statistical thermodynamics developed by Einstein, Boltzmann, and Gibbs. Stanislaw Ulam, a mathematician who helped originate the basic principle behind the hydrogen bomb, once quipped, "Ask not what physics can do for biology. Ask what biology can do for physics." Today, the answer is clear: Biology is forcing physicists to develop new experimental and theoretical tools to explore living cells in action.

For physicists, the most fundamental biological question relates to the basic physical principles behind life. How do the laws of physics, far from thermal equilibrium, lead to the spontaneous formation of matter that can self-organize and evolve into ever more complex structures? To answer this, we need to abstract the organizing principles of living systems from the platform of chemistry underlying the biology—and thus perhaps show that life on Earth is not a miraculous, chance event but an inevitable consequence of the laws of physics. Understanding why life occurs at all would let us predict confidently that life exists elsewhere and perhaps even how it could be detected.

This is important because of another discovery, which appeared online in the scientific journal *Icarus* with the title "Enceladus's measured physical libration requires a global subsurface ocean," by P. C. Thomas, *et al*. This, too, is a coming-of-age story, and recounts a triumph of human ingenuity. NASA sent a spacecraft to Saturn and for seven years it observed with exquisite accuracy the rotation of the moon Enceladus. Enceladus wobbles as it rotates. You probably know that if you have two eggs, one hard-boiled and the other not, you can tell which is which by spinning them and seeing what happens when you stop (try it!).

The big news is that Enceladus is like the raw egg. It wobbles as if filled with liquid. There's a worldwide ocean of water under its surface of solid ice—an ocean presumably kept above freezing by tidal friction and geothermal activity. Enceladus is one place in the solar system where we know there is a large body of warm water and geothermal activity, potentially capable of supporting life as we know it.

The same wonderful spacecraft photographed fountains of water and vapor spurting from Enceladus's south pole and has flown through them to see what molecules are present. Future missions to the Fountains of Enceladus will look specifically for life. I hope Q-Bio will be there too, at least in spirit, predicting what to look for given the moon's geochemistry. And perhaps even predicting that we should confidently expect life everywhere we look.

MATHEMATICS AND REALITY

CLIFFORD PICKOVER

Author, trilogy: *The Math Book, The Physics Book, The Medical Book*

A recent headline in the journal *Nature* declared "PARADOX AT THE HEART OF MATHEMATICS MAKES PHYSICS PROBLEM UNANSWER- ABLE." *3 Quarks Daily* weighed in with "GÖDEL'S INCOMPLETE- NESS THEOREMS ARE CONNECTED TO UNSOLVABLE CALCULATIONS IN QUANTUM PHYSICS." Indeed, the degree to which mathematics describes, constrains, or makes predictions about reality is sure to be a fertile and important discussion topic for years or even centuries to come.

In 1931, mathematician Kurt Gödel determined that some statements are undecidable, suggesting that it is impossible to prove them either true or false. In his first incompleteness the- orem, Gödel recognized that there will always be statements about the natural numbers that are true but unprovable within the system. We now leap forward more than eighty years and learn that Gödel's principle appears to make it impossible to calculate an important property of a material—namely, the gaps between the lowest energy levels of its electrons. Although this finding seems to concern an idealized model of the atoms in a material, some quantum information theorists, such as Toby Cubitt, suggest that this finding limits the extent to which we can predict the behavior of certain real materials and particles.

Prior to this finding, mathematicians also discovered unlikely connections between prime numbers and quantum physics. For example, in 1972, physicist Freeman Dyson and number theorist

Hugh Montgomery discovered that if we examine a strip of zeros from Riemann's critical line in the zeta function, certain experimentally recorded energy levels in the nucleus of a large atom have a mysterious correspondence to the distribution of zeros, which, in turn, has a relationship to the distribution of prime numbers.

Of course, there is a great debate as to whether mathematics is a reliable path to the truth about the universe and reality. Some suggest that mathematics is essentially a product of the human imagination and we simply shape it to describe reality.

Nevertheless, mathematical theories have sometimes been used to predict phenomena that were not confirmed until years later. Maxwell's equations, for example, predicted radio waves. Einstein's field equations suggested that gravity would bend light and that the universe is expanding. Physicist Paul Dirac once noted that the abstract mathematics we study now gives us a glimpse of physics in the future. In fact, his equations predicted the existence of antimatter, which was subsequently discovered. Similarly, mathematician Nikolai Lobachevsky said that "there is no branch of mathematics, however abstract, which may not someday be applied to the phenomena of the real world."

Mathematics is often in the news, particularly as physicists and cosmologists make spectacular advances, even contemplating the universe as a wave function and speculating on the existence of multiple universes. Because the questions that mathematics touches on can be quite deep, we will continue to discuss the implications of the relationship between mathematics and reality perhaps for as long as humankind exists.

SYNTHETIC LEARNING

KEVIN KELLY

Senior Maverick and cofounder, *Wired*; author, *The Inevitable: Understanding the 12 Technological Forces That Will Shape Our Future*

Researchers at DeepMind, an AI company in London, recently reported that they taught a computer system how to learn to play forty-nine simple video games—not "how to play video games" but how to *learn* to play the games. This is a profound difference. Playing a video game, even one as simple as the 1970s classic game *Pong*, requires a suite of sophisticated perception, anticipation, and cognitive skills. A dozen years ago, no algorithms could perform these tasks, but today these game-playing codes are embedded in most computer games. When you play a 2015 video game, you're usually playing against refined algorithms crafted by genius human coders. But rather than program this new set of algorithms to play a game, the DeepMind AI team programmed their machine to learn how to play. The algorithm (a deep neural network) started out with no success in the game and no skill or strategy and then assembled its own code as it played the game, by being rewarded for improving. The technical term is "unsupervised learning." By the end of hundreds of rounds, the neural net could play the game as well as human players, sometimes better.

This learning should not be equated with human intelligence. The mechanics of its learning are vastly different from how we learn. It won't displace humans or take over the world. However, this kind of synthetic learning will grow in capabilities. The significant news is that learning—real unsupervised learning—can be synthesized. Once learning can be synthesized, it can be distributed into all kinds of ordinary devices and functions. It can enable self-driving cars to get better or medical diagnosing programs to improve with use.

Learning, like many other attributes we thought only humans owned, turns out to be something we can program machines to do. Learning can be automated. While simple second-order learning (learning how to learn) was once rare and precious, it will now become routine and common. Just like tireless powerful motors and speedy communications a century ago, learning will quickly become the norm in our built world. All kinds of simple things will learn. Automated synthetic learning won't make your oven as smart as you are, but it will make better bread.

Very soon, smart things won't be enough. Now that we know how to synthesize learning, we'll expect all things to automatically improve as they're used, just as DeepMind's game learner did. Our surprise in years to come will be in the many unlikely places we'll be able to implant synthetic learning.

A GENUINE SCIENCE OF LEARNING

KEITH DEVLIN

Mathematician; executive director, H-STAR Institute, Stanford University; author, *The Man of Numbers: Fibonacci's Arithmetic Revolution*

The education field today is much like medicine was in the 19th century—a human practice guided by intuition, experience, and occasionally inspiration. It took the development of modern biology and biochemistry in the early part of the 20th century to provide the solid underpinnings of today's science of medicine.

To me—a mathematician who became interested in mathematics education in the second half of my career—it seems we may at last be seeing the emergence of a genuine science of learning. Given the huge significance of education in human society, that would make it one of the most interesting and important of today's science stories.

At the risk of raising the ire of many researchers, I should note that I am not basing my assessment on the rapid growth in educational neuroscience—you know, the kind of study where a subject is slid into an fMRI machine and asked to solve math puzzles. Those studies are valuable, but at the present stage, at best, they provide tentative clues about how people learn and little specific in terms of how to help people learn. (A good analogy would be trying to diagnose an engine fault in a car by moving a thermometer over the hood.) Someday educational neuroscience may provide a solid basis for education the way, say, the modern theory of genetics advanced medical practice. But not yet. Rather, the science of learning emerges from the possibilities Internet technology brings to the familiar experimental cognitive-science approach.

The problem that has traditionally beset learning research has been its essential dependence on the individual teacher, which makes it near impossible to run the kinds of large-scale, control-group, intervention studies common in medicine. Classroom studies invariably end up as studies of the teacher as much as of the students, and often measure the effect of the students' home environment rather than what goes on in the classroom.

For instance, news articles often cite the large number of successful people who as children attended a Montessori school, a figure hugely disproportionate to the relatively small number of such schools. Now, it may well be that Montessori educational principles are good ones, but it's also true that such schools are magnets for passionate, dedicated teachers, and the pupils who attend them do so because they have parents who decide to enroll their offspring in such a school and have already raised their children in a learning-rich home environment.

Internet technology offers an opportunity to carry out medical-research-like, large-scale control-group studies of classroom learning which can significantly mitigate the teacher effect and home effect, allowing useful investigation of different educational techniques. Provided you collect the right data, Big Data techniques can detect patterns that cut across the wide range of teacher/teacher and family/family variation, allowing useful conclusions to be drawn.

An important factor is that a sufficiently significant part of the actual learning is done in a digital environment, where every action can be captured. This is not easily achieved. The vast majority of educational software products operate around the edges of learning: providing the learner with information; asking questions and capturing their answers (in a machine-actionable, multiple-choice format); and handling course logistics with a learning management system.

What is missing is any insight into what is actually going on in the student's mind—something that can be very different from what the evidence shows, as was illustrated for mathematics learning several decades ago by a study now famously referred to as "Benny's Rules," where a child who had aced a progressive battery of programmed learning cycles was found to have constructed an elaborate internal, rule-based "mathematics" enabling him to pass all the tests with flying colors, but which was completely false and bore no relation to actual mathematics.

But realtime, interactive software allows for much more than we have seen flooding out of such tech hotbeds as Silicon Valley. To date, some of the more effective uses from the viewpoint of running large-scale, comparative-learning studies have been by way of learning video games—so-called game-based learning. (It remains an open question how significant the game element is in terms of learning outcomes.)

In elementary through middle-school mathematics learning (the research I am familiar with), what has been discovered, by a number of teams, is that digital learning interventions of as little as 10 minutes a day can, in as little as a month, result in significant learning gains when measured by a standardized test—with improvements of as much as 20 percent in some key thinking skills. That may sound like an educational magic pill; it almost certainly is not. It's most likely an early sign that we know even less about learning than we thought we did.

Part of what's going on is that many earlier studies measured knowledge rather than thinking ability. The learning gains found in the studies I refer to are not knowledge acquired or algorithmic procedures mastered but high-level problem-solving ability. What's exciting about these findings is that in today's information- and computation-rich environment those human problem-solving skills are now at a premium.

Like any good science, and in particular any new science, this work has generated far more research questions than it has answered. Indeed, it is too early to say whether it has answered *any* questions. Rather, as of now we have a scientifically sound method to conduct experiments at scale, some suggestive early results, and a long and growing list of research questions—all testable. Looks to me like we're about to see the emergence of a genuine *science* of learning.

BAYESIAN PROGRAM LEARNING

JOHN C. MATHER

Senior astrophysicist, Observational Cosmology Laboratory, NASA's Goddard Space Flight Center; recipient, 2006 Nobel Prize in physics; co-author (with John Boslough), *The Very First Light*

You may not like it! But artificial intelligence jumped a bit closer in 2015 with the development of Bayesian program learning, described by Lake, Salakhutdinov, and Tenenbaum in *Science* ("Human-level concept learning through probabilistic program induction"). It's news because for decades I've been hearing about how hard it is to achieve artificial intelligence, and the most successful methods have used brute force. Methods based on understanding the symbols and logic of things and language have had a tough time. The challenge is to invent a computer representation of complex information and then enable a machine to learn that information from examples and evidence.

Lake *et al.* give a mathematical framework, an algorithm, and a computer code that implements it, and their software has learned to read 1,623 handwritten characters in fifty languages as well as a human being. They write: "Concepts are represented as simple probabilistic programs—that is, probabilistic generative models expressed as structured procedures in an abstract description language." Also, a concept can be built up by re-using parts of other concepts or programs. The probabilistic approach handles the imprecision both of definitions and examples. (Bayes' theorem tells us how to compute the probability of something complicated if we know the probabilities of various smaller things that go into the complicated thing.) Their system can learn very quickly, sometimes in one shot or from a few examples, in a humanlike way, and with humanlike accuracy. This ability is in dramatic contrast to competing methods depending on immense data sets and simulated neural networks, which are always in the news.

So now there are many new questions: How general is this approach? How much structure do humans have to give it to get it started? Could it really be superior in the end? Is this how living intelligent systems work? How could we tell? Can this computer system grow enough to represent complex concepts important to humans in daily life? Where are the first practical applications?

This is a long-term project, without any obvious limits to how far it can go. Could this method be so efficient that it doesn't take a super-duper supercomputer to achieve, or at least represent, artificial intelligence? Insects do very well with a tiny brain, after all. More generally, when do we get accurate transcriptions of multi-person conversations, instantaneous machine-language translation, scene recognition, face recognition, self-driving cars, self-directed drones safely delivering packages, machine understanding of physics and engineering, machine representation of biological concepts, and machine ability to read the Library of Congress and discuss it in a philosophy or history class? When will my digital assistant really understand what I want to do? Is this how the intelligent Mars rover will hunt for signs of life on Mars? How about military offense and defense? How could this system implement Asimov's three laws of robotics to protect humans from robots?

How would you know whether to trust your robot? When will people be obsolete?

I'm sure many people are already working on all these questions. I see opportunities for mischief, but the defense against the dark arts will push rapid progress, too. I am both thrilled and frightened.

FSM (FECES-STANDARD MONEY)

JAEWEON CHO

Professor, Environmental Engineering, UNIST, Ulsan, Republic of Korea

We are facing problems from two of our great inventions: money and flushing the toilet.

We all use money, but we can be isolated from money at the same time. Money is one of the greatest human inventions, but it may also be among the worst ever created.

Our present monetary system has nothing to do with anything that comes from human beings. While we can do many things with money in our modern societies, there are no significant connections between the money and ourselves. Thus, it can be hypothesized that whenever we use money, we are isolating ourselves from the world.

Flushing the toilet—a second great invention—also has both positive and negative aspects. While it deals effectively with issues of hygiene, when we flush the toilet we are flushing our excretion into the natural environment, and this leads to severe problems.

Here's an idea that could lead us into a new kind of fiscal world: Can you think of how we might mitigate the problems while keeping the advantages of our current money system? Imagine a scientific method of making odorless powder from our feces and replacing money with that powder: that is, feces-standard money, or FSM.

Every morning we can put our powder into reactors situated in our communities to supply food for the microorganisms that can produce various fuels, such as methane and biodiesel. We can receive a certain amount of FSM in exchange for the powdered feces and use the FSM to obtain any equivalent value within a system. Feces, like ordinary currency, is limited and precious; nobody can make more than a certain limit, and it can be converted to energy.

Furthermore, everyone can make feces every day. Whenever we produce and use FSM, it will remind us of the bottom-line connection between FSM and human beings. Thus, FSM has meaning from the perspectives both of the economy and the human mind.

FSM can become "basic income" as long as we put feces into the reactors on a daily basis instead of continuing to flush the toilet. We don't have to rank the feasibility of the concept of basic income against that of our present monetary system. We can use both money systems without conflicts.

FSM is different from other types of credit, such as mileage, coupons, and online coins, because it is directly connected to our existence and free will—our intention to not use the flushable toilet.

Suggestions: An FSM system can be designed with an app or in other, similar ways. FSM can be used, along with the present currency, to buy such things as gas, coffee, food, and to pay for various enjoyable activities. Values depending on the production of energy or other equivalent products from the feces at a designated time or date can be distributed to the participants of the system. (Of course, since FSM can't provide everything we need, conventional money also has to be used.) The proposed money system will also require development of various technologies, such as biological processes for energy production—and new industries as well, such as the manufacture of bathroom appliances for the conversion of feces into powder.

THE IRONIES OF HIGHER ARITHMETIC

JIM HOLT

Philosopher and essayist; author, *Why Does the World Exist?*

The *abc* conjecture, first proposed in 1985, asserts a surprising connection between the addition and multiplication of whole numbers. (The name comes from that amiable equation, $a + b = c$.) It appears to be one of the deepest and most far-reaching unresolved conjectures in mathematics, intimately tied up with Roth's theorem, the Mordell conjecture, and the generalized Szpiro conjecture.

In 2012, Shinichi Mochizuki of Kyoto University claimed to have proved the *abc* conjecture—potentially a stunning advance in higher mathematics. But is the proof sound? No one has a clue. Near the end of last year, some of the world's leading experts on number theory convened in Oxford to sort *abc* out. They failed.

Mochizuki's would-be proof of the *abc* conjecture uses a formalism he calls "inter-universal Teichmüller theory" (IUT), which features highly symmetric algebraic structures dubbed Frobenioids. At first, the problem was that no one (except, we must suppose, its creator) could understand this new and transcendently abstract formalism. Nor could anyone see how it might bear on *abc*. By the time of the Oxford gathering, however, three mathematicians—two of them colleagues of Mochizuki's at Kyoto, the third from Purdue in the U.S.—had seen the light. But when they tried to explain IUT and Frobenioids, their peers had no idea what they were talking about—"indigestible," one of the participants called their lectures.

In principle, checking a proof in mathematics shouldn't require any intelligence or insight. It's something a machine could do. In practice, though, a mathematician never writes out the sort of austerely detailed "formal" proof that a computer might check. Life is too short. Instead, she (lady friend of mine) offers a more or less elaborate argument that such a formal proof exists—an argument that, she hopes, will persuade her peers.

With IUT and *abc*, this business of persuasion has got off to a shaky start. So far, converts to the church of Mochizuki seem incapable of sharing their newfound enlightenment with the uninitiated. The *abc* conjecture remains a conjecture, not a theorem. That might change this summer, when number theorists plan to reconvene, this time in Kyoto, to struggle anew with the alleged proof.

So what's the news? It's that mathematics—which in my cynical moods I tend to regard as little more than a giant tautology, one that would be as boring to a transhuman intelligence as tic-tac-toe is to us—is really something weirder, messier, more fallible, and far more noble.

And that "Frobenioids" is available as a name for a Brooklyn indie band.

BROKE PEOPLE IGNORING $20
BILLS ON THE SIDEWALK

MICHAEL VASSAR

Cofounder and chief science officer, MetaMed Research

It's not every year that *Edge* echoes *South Park*. I guess everyone's trying to figure out what's real right now. The 2015 *South
Park* season revolved around people losing the ability to distinguish between news and advertising. One day they wake up
broke, at war, and unable to easily distinguish friend from foe.
News, as a concept, is gone. Science, as a concept, is gone. In
information warfare, the assumption that reliable, low-context
communication is even possible recedes into fantasy, taking
with it both news and science and replacing them with politics
and marketing. I think the real news, viewed from behind the
new extra-strength Veil of Maya, is what you think you see
with your own eyes and have checked out against the analytic
parts of the scholarly literature. Here's what I've got.

Last November, I was visiting Universidad Francisco Mar-
roquín, in Guatemala, which is known primarily for being the
most libertarian university in the world. While there, on the floor
of the Economics Department my co-conspirator found local
currency worth almost exactly US$20. Technically, the money
wasn't visible from the sidewalk, but the signs announcing gasoline prices clearly are. In the last few years, I have observed
those prices decoupling both from the price of oil and from one
another, whether across town or even across the street. Growing
up, I always noticed whether the gas prices on opposite sides of

the road differed by 1, 2, or sometimes even 3 cents. Today, such prices typically differ by more than $.20. I recently saw two stations across the street from one another with prices differing by $.36/gallon and two stations a mile apart charging respectively $2.49/gallon and $3.86/gallon. For a median American driver, $.20/gallon, invested at historically normal rates of return, would add up to about $1,500 over the next decade. Median retirement savings for families aged 55–64 is only $15,000, and for families with retirement accounts, median savings are still only $150,000.

I'm OK with people not behaving like *Homo economicus*, but if broke people are becoming less economically rational with time, this suggests that people don't feel that they can predict the future in basic respects. That they aren't relying on savings to provide for their basic needs. Who can blame them for financial recklessness? Theory, as well as practice, tells us that their leaders have been setting an ever worse example for generations.

Financial economics provides the analytic literature on economic caution and risk. In 1987, Larry Summers and Brad DeLong showed that given a risk premium (a standard assumption in financial economics), irrational noise traders crowd out rational actors over time ("The Economic Consequences of Noise Traders"). When Peter Thiel talks about the shift from "concrete optimism" to "abstract optimism," he's characterizing the pattern selected for by this dynamic. This shift toward noise trading inflates equity prices, concentrates wealth, and causes more speculative assets to command higher prices each decade than similarly speculative assets would have commanded in the previous decade. With 84 percent of corporate valuations now taking the form of intangibles, up from 16 percent forty years ago—that sounds like the world I see around me. The overall divergence of map from territory in economic settings ultimately means the annihilation of strategy as we know it for most people, making economic prudence a predictably losing strategy, which means that in the long run if we can't figure out a better way to aggregate local economic information, we won't have the patience to effectively use that information.

WE FEAR THE WRONG THINGS

DAVID G. MYERS

Professor of psychology, Hope College; author, *Intuition: Its Powers and Perils*

If we knew that AK-47-wielding terrorists would kill 1,000 people in the U.S. in 2016, a thinkable possibility, then we should be afraid—albeit 1/10 as afraid of other homicidal gun violence (which kills more than 10,000), and 1/20 as fearful of riding in a motor vehicle, where 22,000 Americans die annually. Yet several recent surveys show us to be much less fearful of the greater everyday threats than of the dreaded horror. The hijacking of our rationality by fears of terrorist guns highlights an important and enduring piece of scientific news: *We often fear the wrong things.*

Shortly after 9/11, when America was besieged by fear, I offered a calculation: If we now flew 20 percent less and instead drove half those unflown miles, then, given the greater safety of scheduled airline flights, we could expect about 800 more people to die on our roads. Why do we fear flying, when, for most of us, the most dangerous part of our trip is the drive to the airport? Why do terrorist fears so effectively inflate our stereotypes of Muslims, inflame "us vs. them" thinking, and make many of those of us who are Christians forget the ethics of Jesus ("I was a stranger and you welcomed me.")?

Underlying our exaggerated fears is the "availability heuristic": We fear what's readily available in memory. Vivid, cognitively available images—a horrific air crash, a mass slaughter—distort our judgments of risk. Thus we remember—

and fear—disasters (tornadoes, plane crashes, attacks) that kill people dramatically, in bunches, whereas we fear too little the threats that claim lives one by one. Bill Gates once observed that we hardly notice the half-million children quietly dying each year from rotavirus—the equivalent of four 747s full of children *every day*. And we discount the future (and its future weapon of mass destruction, climate change). If only such deaths were more dramatic and immediate. Imagine (to adapt one mathematician's suggestion) that cigarettes were harmless—except, once in every 25,000 packs, for a single cigarette filled with dynamite. Not such a bad risk of having your head blown off. But with 250 million packs a day consumed worldwide, we could expect more than 10,000 gruesome daily deaths (the approximate actual toll of cigarette smoking)—surely enough to have cigarettes banned.

News-fed images can make us excessively fearful of infinitesimal risks. And so we spend an estimated $500 million on antiterrorism security per U.S. terrorist death but only $10,000 on cancer research per cancer death. As one risk expert explained, "If it's in the news, don't worry about it. The very definition of *news* is 'something that hardly ever happens.'"

It's entirely normal to fear violence from those who despise us. But it's also smart to be mindful of the realities of how most people die, lest the terrorists successfully manipulate our politics. With death on their minds, people exhibit "terror management." They respond to death reminders by derogating those who challenge their worldviews. Before the 2004 election, reported one research team, reminders of 9/11 shifted people's sympathies toward conservative politicians and antiterrorism policies.

Media researcher George Gerbner's cautionary words to a 1981 congressional subcommittee ring true today: "Fearful people are more dependent, more easily manipulated and controlled, more susceptible to deceptively simple, strong, tough measures and hard-line postures."

Ergo, we too often fear the wrong things. And it matters.

LIVING IN TERROR OF TERRORISM

GERD GIGERENZER

Psychologist; director, Center for Adaptive Behavior and Cognition, Max Planck Institute for Human Development, Berlin; author, *Risk Savvy*

Terrorism has indeed caused a huge death toll in countries such as Afghanistan, Syria, and Nigeria. But in Europe or North America, a terrorist attack is not what will likely kill you. In a typical year, more Americans die from lightning than terrorism. A great many more die from second-hand smoke and "regular" gun violence. Even more likely, Americans can expect to lose their lives from preventable medical errors in hospitals, even in the best of them. The estimated number of unnecessary deaths has soared from 98,000 in 1999 to 440,000 annually, according to a recent study in the *Journal of Patient Safety*.

Why are we scared of what most likely won't kill us? Psychology provides an answer. It is called "fear of dread risks." This fear is elicited by a situation in which many people die *within a short time*. Note that the fear is not about dying but about suddenly dying together with many others at one point in time. When as many—or more—people die distributed over the year, whether from gun violence, motorcycle accidents, or in hospital beds, it's hard to conjure up anxiety.

For that reason terrorists strike twice. First with physical force and second by capitalizing on our propensity for dread-risk fear. After 9/11, many Americans avoided flying and used their cars instead. As a consequence, some 1,600 people died

348

from the resulting traffic accidents, more than the total number of individuals killed aboard the four hijacked planes. That can be called Osama bin Laden's second strike. All those people could still be alive if they had flown instead of driven, seeing as there was no deadly accident on commercial airline flights in the U.S. for a number of years thereafter.

Although billions have been poured into Homeland Security and similar institutions to prevent the first strike of terrorists, almost no funding has been provided to prevent the second strike. I believe that making the public psychologically aware of how terrorists exploit our fears could save more lives than NSA Big Data analytics. It could also open people's eyes to the fact that some politicians and other interest groups work on keeping our dread-risk fear aflame to nudge us into accepting personal surveillance and restriction of our democratic liberties. Living in terror of terrorism can be more dangerous than terrorism itself.

THE STATE OF THE WORLD ISN'T AS BAD AS YOU THINK

STEVEN R. QUARTZ

Neuroscientist; professor of philosophy, Caltech; co-author (with Anette Asp), *Cool*

If you find yourself at a cocktail party searching for a conversation starter, I'd recommend the opening line of a recent Bill and Melinda Gates Annual Letter: "By almost any measure, the world is better than it has ever been." Although people will react with incredulity at the very possibility that things could be getting better, they'll welcome the opportunity to straighten you out. Be prepared for the recitation of the daily headlines—bad news piled on top of worse—that will inevitably follow. Virtually everyone I've mentioned this quote to is sure it's wrong.

For example, about two-thirds of Americans believe that the number of people living in extreme poverty has doubled in the last twenty years. People point to conflicts in the Middle East as evidence of a world in chaos, the retreat of democracy, plummeting human rights, and a global decline of well-being. Yet the news from social science through the accumulation of large-scale, longitudinal data sets belies this declinist worldview.

In reality, extreme poverty has nearly halved in the last twenty years—about a billion people have escaped it. Material well-being—income, reduced infant mortality, life expectancy, educational access (particularly for females)—has grown at its greatest pace in the last few decades. The number of democracies among the developing nations has tripled since the 1980s, while the number of people killed in armed conflicts has decreased by 75 percent. This isn't the place to delve into the details of how large-scale statistical data sets, and ones increasingly representative of the world's population, provide a more accurate, though deeply counterintuitive, assessment of the state of the world (for that, see Steven Pinker's *The Better Angels of our Nature*).

Instead, I want to suggest three reasons why it's such important scientific news. First, although these long-term trends may not resuscitate an old-fashioned notion of progress—certainly not one suggesting that history possesses intrinsic directionality—they do call out for a better understanding (and recognition) of the technological and cultural dynamics driving long-term patterns of historical change. What's even more interesting is their stark demonstration of how deeply our cognitive and emotional biases distort our worldview. In particular, there's good evidence that we don't remember the past as it was. Instead, we systematically edit it, typically omitting the bad and highlighting the good, leading to cognitive biases such as "rosy retrospection."

At the cultural level, these biases make us vulnerable to declinist narratives. From Pope Francis's anti-modernist encyclical to capitalism's inevitable death by internal contradictions and moral decline, declinist narratives intuitively resonate with our cognitive biases. They are thus an easy sell and cause us to lose sight of the fact that until a few centuries ago the world's population was stuck in abject poverty, a subsistence-level Malthusian trap of dreary cycles of population growth and famine.

In reality, not only has material well-being increased around the globe but global inequality is also decreasing, as a result of technological and cultural innovations driving globalization. We should be particularly on guard against declinist narratives that also trigger our emotional biases, especially those hijacking the brain's low-level threat-detection circuits. These alarmist narratives identify an immediate or imminent threat, a harbinger of decline, which unconsciously triggers the amygdala and initiates a cascade of brain chemicals—norepinephrine, acetylcholine, dopamine, serotonin, and the hormones adrenalin and cortisol—creating primal, visceral feelings of dread and locking our attention to that narrative, effectively shutting down rational appraisal.

Much of what counts as "news" today involves such narratives. The combination of an ever-shortening news cycle, near-instantaneous communications, fragmented markets, heightened competition for viewership, and our cognitive and emotional biases conspire to make it all but inevitable that these narratives dominate—and unlikely for us to grasp the progressive themes that large-scale data analyses reveal.

The result is today's dominant alarmist and declinist news cycle, which is essentially a random walk from moral panic to moral panic. To appreciate the real news—that by many fundamental measures the state of the world is improving—requires an exercise in cognitive control, inhibiting our first emotional impulses and allowing a rational appraisal of scientifically informed data. This is by no means some Pollyanna-ish exercise of denial. The most important scientific news to me is that the broad historical trajectory of human societies provides a powerful counternarrative to today's dominant declinist worldview.

THE HEALTHY DIET U-TURN

ED REGIS

Science writer; author, *Monsters: The Hindenburg Disaster and the Birth of Pathological Technology*

To me, the most interesting bit of news in the last couple of years was the sea change in attitude among nutritional scientists—from promotion of an anti-fat, pro-carbohydrate set of dietary recommendations to a lower-carbohydrate, selectively pro-fat dietary regime. The issue is important, because human health and, indeed, human lives are at stake.

For years Americans had been told by the experts to avoid fats at all costs, as if fats were the antichrists of nutrition. A diet low in fats and rich in carbohydrates was the way to go in order to achieve a sleek, gazelle-like body and physiological enlightenment. In consequence, no-fat or low-fat foods became all the rage; for a long time, the only kind of yogurt you could find on grocery shelves was the jelly-like zero-fat variety and the only available canned tuna was packed not in olive oil but in water, as if the poor creature were still swimming.

Unappetizing as much of it was, many Americans duly followed this stringent set of dietary dos and don'ts. But we did not thereby become a nation of fit, trim, and healthy physical specimens—far from it. Instead, we suffered an obesity epidemic across all age groups, a tidal wave of heart disease, and highly increased rates of Type 2 diabetes. Once they were digested, all those carbohydrate-rich foods got converted into glucose, which raised insulin levels and in turn caused storage of excess bodily fat.

Nutritional scientists learned the dual lesson that a diet high in carbohydrates can in fact be hazardous to your health, and that their fat phobia was unjustified by the evidence. In reality, there are good fats (like olive oil) and bad fats, healthy carbs and unhealthy carbs (like refined sugars). Many nutritionists now favor a diametrically opposite approach, allowing certain fats as wholesome and healthy, while calling for a reduction in carbohydrates, especially refined sugars and starches.

A corollary of this about-face in dietary wisdom was the realization that much of so-called nutritional "science" was bad science to begin with. Many of the canonical studies of diet and nutrition were flawed by selective use of evidence, unrepresentative sampling, absence of adequate controls, and shifting clinical-trial populations. Furthermore, some of the principal investigators were prone to selection bias and loath to refute their preconceived viewpoints with contrary evidence. (These and other failings of the discipline are exhaustively documented in journalist Nina Teicholz's book *The Big Fat Surprise* [2014].)

Unfortunately, nutritional science remains something of a backwater. NASA's *Curiosity* rover explores the plains, craters, and dunes of Mars, and the *New Horizons* spacecraft takes exquisite pictures of Pluto. Molecular biologists wield superb gene-editing tools and are in the process of resurrecting extinct species. Nevertheless, when it comes to the prosaic task of telling us what to eat to achieve good health and avoid heart disease, obesity, and other ailments, dietary science still has a long way to go.

FATTY FOODS ARE GOOD FOR YOUR HEALTH

PETER TURCHIN

Professor, Department of Ecology and Evolutionary Biology,
University of Connecticut; author, *Ultrasociety: How 10,000 Years of
War Made Humans the Greatest Cooperators on Earth*

Amid all the confusing fluctuations in dietary fashion to which
Americans have been exposed since the 1960s, one recommendation has remained unchallenged. Beginning in the 1960s,
and until 2015, Americans have been getting consistent dietary
advice: Fat, especially saturated fat, is bad for your health. By
the 1980s, the belief equating a low-fat diet with better health
had become enshrined in the national dietary advice from the
U.S. Department of Agriculture and was endorsed by the surgeon general. Meanwhile, as Americans ate less fat, they steadily
became more obese.

The obesity epidemic probably has many causes, not all well
understood. But clearly the misguided dietary advice bombard-
ing us over the past five decades is an important contributing
factor.

There has in fact never been any scientific evidence that
cutting down total fat consumption has a positive effect on
health—specifically, reduced risks of heart disease and diabetes.
For years, those who pointed this out were marginalized, but
recently evidence debunking the supposed benefits of low-fat
diets has reached a critical mass, so that a mainstream magazine
such as *Time* could write in 2014: "SCIENTISTS LABELED FAT THE

ENEMY. WHY THEY WERE WRONG." And now the official *Scientific Report of the 2015 Dietary Guidelines Advisory Committee* admits that much.

There are several reasons why eating a low-fat diet is bad for your health. One is that if you lower the proportion of fat in your diet, you must replace it with something else. Eating more carbohydrates (whether refined or "complex") increases your chances of becoming diabetic. Eating more proteins increases your chances of getting gout.

But perhaps a more important reason is that many Americans stopped eating *food* and switched to highly-processed food substitutes: margarine, processed meats (such as the original Spam—not to be confused with email spam), low-fat cookies, and so on. In each case, we now have abundant evidence that these are "anti-health foods," because they contain artificial trans fats, preservatives, or highly processed carbohydrates.

While controlled diet studies are important and necessary for making informed decisions about our diets, an exciting recent scientific breakthrough has resulted from the infusion of evolutionary science into nutrition science. After all, you need to figure out what hypotheses you want to test with controlled trials, and evolution turned out to be a fertile generator of theoretical ideas for such tests.

One of the sources of ideas to test clinically is the growing knowledge of the characteristic diets of early human beings. Consider this simple idea (although it clearly was too much for traditional nutritionists): We will be better adapted to something eaten by our ancestors over millions of years than to, say, margarine, which we first encountered only 100 years ago. Or take a food like wheat, to which some populations (those in the Fertile Crescent) have been exposed for 10,000 years and others

(Pacific Islanders) for only 200 years. Is it surprising that Pacific Islanders have the greatest prevalence of obesity in the world, higher even than in the United States? And should we really tell them to switch to a Mediterranean diet, heavy on grains, pulse, and dairy, to which they've had no evolutionary exposure whatsoever?

Our knowledge of ancestral diets is growing rapidly. We're adapted to eating a variety of fatty foods, including grass-fed ruminants (beef and lamb) and seafood (oily fish), both good sources of Omega-3 fatty acids. Of particular importance could be bone marrow—it's likely that the first members of the genus *Homo* (e.g., *habilis*) were not hunters but scavengers, who competed with hyenas for large marrow bones. It's also likely that nutrients from bone marrow (and brains!) of scavenged savannah ungulates were the key resource for the evolution of our oversized brains.

The new knowledge explains why Americans are getting fatter by eating low-fat diets. When you eliminate fatty foods that your body (especially your brain) needs, your body starts sending you persistent signals that you're malnourished. So you'll overeat foods other than fatty ones. The extra, unnecessary calories you consume (probably from carbohydrates) will be stored as fat. As a result, you'll be unhappy, unhealthy, and overweight. You can avoid those extra pounds, of course, if you have a steely will (which few people have). Then you won't be overweight— just unhappy and unhealthy.

So, to lose fat you need to eat, not fat, but fatty *foods*. Paradoxically, eating enough fatty food of the right sorts will help make you lean, as well as happy and—*Edge* readers, take note— smart!

PARTISAN HOSTILITY

JONATHAN HAIDT

Social psychologist; professor, NYU-Stern School of Business; author,
The Righteous Mind

If you were on a committee tasked with choosing someone to
hire (or admit to your university, or receive a prize in your field)
and it came down to two candidates who were equally quali-
fied on objective measures, which would you be most likely to
choose?

(a) The one who shared your race?
(b) The one who shared your gender?
(c) The one who shared your religion?
(d) The one who shared your political party or ideology?

The correct answer, for most Americans, is now (d). It is surely
good news that prejudice based on race, gender, and religion are
down in recent decades. But it is bad news—for America, for the
world, and for science—that partisan hostility is way up.

A 2014 paper by Princeton University political scientists
Shanto Iyengar and Sean Westwood, titled "Fear and Loathing
Across Party Lines: New Evidence on Group Polarization," reports four studies (all using nationally representative samples) in
which respondents were given various ways to reveal both cross-
partisan and cross-racial prejudice, and in all cases cross-partisan
prejudice was larger.

Iyengar and Westwood used a measure of implicit attitudes

(called the Implicit Association Test), which gauges how quickly and easily people pair words with "good" and "bad" connotations with words and images associated with "African Americans" vis-à-vis "European Americans." They also ran a new version of the test using words and images related to Republicans and Democrats. The effect sizes for cross-partisan implicit attitudes were much larger than those for race. For white participants who identified with a party, the cross-partisan effect was about 50 percent larger than the cross-race effect. That is, when Americans look at or listen to one another, their automatic associations are more negative toward people from the "other side" than toward people of a different race.

In another study, they had participants read pairs of fabricated résumés of graduating high school seniors and select one to receive a scholarship. Race made a difference—black and white participants generally preferred to award the scholarship to the student with the stereotypically black name. But party affiliation made an even bigger difference, and always in a tribal way: About 80 percent of the time, participants selected the candidate belonging to their party, and it made little difference whether their co-partisan had a higher or lower GPA than the other candidate.

In the two further studies, Iyengar and Westwood had participants play behavioral economics games (the "trust game" and the "dictator game"). Each person played with what they thought was a particular other person, about whom they read a brief profile including the person's age, gender, income, race, and political ideology. Race and ideology were manipulated systematically. Race made no difference, but partisanship mattered a lot: People were more trusting and generous when they thought they were playing with a co-partisan than an opposing partisan.

This is bad news for America because it is hard to have an effective democracy without compromise. But rising partisan hostility means that Americans increasingly see the other side not just as wrong but as evil, as a threat to the very existence of the nation, according to Pew Research. Americans can expect rising polarization, nastiness, paralysis, and governmental dysfunction for a long time to come.

This is a warning for the rest of the world because some of the trends that have driven America to this point are occurring in many other countries, including rising education and individualism (which make people more ideological), rising immigration and ethnic diversity (which reduces social capital and trust), and stagnant economic growth (which puts people into a zero-sum mindset).

This is bad news for science and universities because universities are usually associated with the left. In the United States, universities have moved rapidly left since 1990, when the left/right ratio of professors across all departments was less than 2:1. By 2004, the left/right ratio was roughly 5:1, and it is still climbing. In the social sciences and humanities, it's far higher. Because this political purification is happening at a time of rising cross-partisan hostility, we can expect increasing hostility from Republican legislators toward universities and the things they desire, including research funding and freedom from federal and state control.

Tribal conflicts and tribal politics took center stage in 2015. Iyengar and Westwood help us understand that tribal conflicts are no longer primarily about race, religion, or nationality. Cross-partisan prejudice should become a focus of concern and further research. In the United States, it may even be a more urgent problem than racial prejudice.

COGNITIVE SCIENCE TRANSFORMS MORAL PHILOSOPHY

STEPHEN P. STICH
Board of Governors Professor, Department of Philosophy, Rutgers University

For 2,500 years, moral philosophy was entrusted to philosophers and theologians. But in recent years moral philosophers who are also cognitive scientists and cognitive scientists with a sophisticated mastery of moral philosophy have transformed moral philosophy. Findings and theories from many branches of cognitive science have been used to reformulate traditional questions and defend substantive views on some of the most important moral issues facing contemporary societies. In this new synthesis, the cognitive sciences are not replacing moral philosophy; rather, they are providing new insights into the psychological and neurological mechanisms underlying moral reasoning and moral judgment, and these insights are being used to construct empirically informed moral theories that are reshaping moral philosophy.

Here's the backstory: From Plato onward, philosophers concerned with morality have made claims about the way the mind works when we consider moral issues. But these claims were always speculative and often set out in metaphors or allegories. With the emergence of scientific psychology in the 20th century, psychologists became increasingly interested in moral judgment and moral development. But much of this work was done by researchers who had little or no acquaintance with the rich philosophical tradition that had drawn important distinctions and defended sophisticated positions about a wide range of moral issues. So philosophers who dipped into this work typically found it naïve and unhelpful.

At the beginning of the current century, that began to change. Prompted by the interdisciplinary zeitgeist, young philosophers (and a few who weren't so young) resolved to master the methods of contemporary psychology and neuroscience and use them to explore questions about the mind which philosophers had been debating for centuries. On the other side of the disciplinary divide, psychologists, neuroscientists, and researchers interested in the evolution of the mind began to engage with the philosophical tradition more seriously. What started as a trickle of papers that were both scientifically and philosophically sophisticated has turned into a flood. Hundreds are published every year, and moral psychology has become a hot topic. There are many examples of this extraordinary work. I'll mention just three.

Joshua Greene is in many ways the poster child for the new synthesis of cognitive science and moral philosophy. While working on his PhD in philosophy, Greene had the altogether novel idea of asking people to make judgments about moral dilemmas while in a brain scanner. Philosophers had already constructed a number of hypothetical moral dilemmas in which a protagonist was required to make a choice between two courses of action. One choice would result in the death of five innocent people; the other would result in the death of one innocent person. But philosophers were puzzled by the fact that in similar cases people sometimes chose to save the five and sometimes chose to let the five die. What Greene found was that different brain regions were involved in these choices. When the five were saved, the brain regions involved were thought to be associated with rational deliberation; when the five were not saved, the brain regions involved were thought to be associated with emotion.

This early result prompted Greene to retrain as a cognitive neuroscientist and triggered a tsunami of studies exploring what goes on in the brain when people make moral judgments. But although Greene became a cognitive scientist, he was still a philosopher, and he draws on a decade of work in moral psychology to defend his account of how moral decisions that divide groups should be made.

If Greene is the poster child for the new synthesis, John Mikhail is its Renaissance man. While completing a philosophy PhD, he spent several years studying cognitive science and then got a law degree. He's now a law professor whose areas of expertise include human-rights law and international law. Drawing on the same family of moral dilemmas that were center stage in Greene's early work, Mikhail has conducted an extensive series of experiments that, he argues, support the view that all normal humans share an important set of innate moral principles. Mikhail argues that this empirical work provides the much needed intellectual underpinning for the doctrine of universal human rights.

And finally an example—one among many—of the new questions that the new synthesis enables us to see. Recent work in psychology reveals that we all have a grab bag of surprising implicit biases. Many people, including people who support and work hard to achieve racial equality, nonetheless associate black faces with negative words and white faces with positive words. And there is a growing body of evidence suggesting that these implicit biases also affect our behavior, though we are usually unaware this is happening.

Moral philosophers have long been concerned to characterize the circumstances under which people are reasonably held to be morally responsible for their actions. Are we morally responsible for behavior influenced by implicit biases? That question has sparked heated debate, and it could not have been asked without the new synthesis.

Will all this still be news in the decades to come? My prediction is that it will. We have only begun to see the profound changes the new synthesis will bring about in moral philosophy.

MORALITY IS MADE OF MEAT

OLIVER SCOTT CURRY

Departmental Lecturer, Institute of Cognitive and Evolutionary
Anthropology, University of Oxford

What is morality and where does it come from? Why does it
exert such a tremendous hold over us? Scholars have struggled
with these questions for millennia, and for many people the
nature of morality is so baffling that they assume it must have a
supernatural origin. But the good news is that we now have a
scientific answer to these questions.

Morality is made of meat. It is a collection of biological and
cultural solutions to the problems of cooperation recurrent in
human social life. Which problems? Caring for families, working in teams, trading favors, resolving conflicts. Which solutions? Love, loyalty, reciprocity, respect. The solutions arose
first as instincts designed by natural selection; later, they were
augmented and extended by human ingenuity and transmitted
as culture. These mechanisms motivate social, cooperative, and
altruistic behavior, and they provide the criteria by which we
evaluate the behavior of others. And why is morality felt to be
so important? Because, for a social species like us, the benefits
of cooperation (and the opportunity costs of its absence) can
hardly be overstated.

The scientific approach was news when Aristotle first hypothesized that morality was a combination of the natural, the
habitual, and the conventional—all of which helped us fulfill
our potential as social animals. It was news when Hobbes the-

orized that morality was an invention designed to corral selfish individuals into mutually beneficial cooperation. It was news when Hume proposed that morality was the product of animal passions and human artifice, aimed at the promotion of the "publick interest." It was news when Darwin conjectured that "the so-called moral sense is aboriginally derived from the social instincts," which tell us how "to act for the public good." And it has been front-page news for the past few decades, as modern science has made discovery after discovery into the empirical basis of morality, delivering evolutionary explanations, animal antecedents, psychological mechanisms, behavioral manifestations, and cultural expressions.

Unfortunately, many philosophers, theologians, and politicians have yet to get the message. They make out that morality is still mysterious, that without God there is no morality, and that the irreligious are unfit for office. This creationist account of morality—"good of the gaps"—is mistaken and alarmist. Morality is natural, not supernatural. We are good because we want to be, and because we are sensitive to the opinions—the praise and the punishment—of others. We can work out for ourselves how best to promote the common good, and with the help of science make the world a better place.

Now, ain't that good news? And ain't it high time we recognized it?

PEOPLE KILL BECAUSE IT'S THE RIGHT THING TO DO

JAMES J. O'DONNELL

Classics scholar; university librarian, Arizona State University; author, *Pagans: The End of Traditional Religion and the Rise of Christianity*

People kill because it's the right thing to do.

In their 2014 book *Virtuous Violence: Hurting and Killing to Create, Sustain, End, and Honor Social Relationships*, moral psychologist Tage Shakti Rai at Northwestern and psychological anthropologist Alan Page Fiske at UCLA sketch the extent to which their work shows that violent behavior among human beings is often not a breach of moral codes but an embodiment of them.

In a sense, we all know this, by way of the exceptions we permit. Augustine's theory of the Just War arose because his god demonstrably approved of some wars. When Joshua fought the battle of Jericho, he had divine approval, "Thou shalt not kill" be damned. To Augustine's credit and others in that tradition, the Just War theory represents hard work to resist as much licit violence as possible. To their discredit, it represents their decision to cave in to questionable evidence and put a stamp of approval on slaughter. (Am I hallucinating by recalling a small woodcut of Augustine in the margin of a *Time* essay on the debates over the justice of the Vietnam War? If my hallucination is correct, I remember shuddering at the sight.)

And certainly we have plenty of examples closer to date: Mideast terrorists and anti-abortion assassins are flamboyant examples, but elected statesmen—Americans as well as those from countries we aren't so fond of—are no less prone to justify killing based on the soundest moral arguments. We glance away nervously and mutter about exceptions. But what if the exceptions are the rule?

If the work of Rai and Fiske wins assent, it points to something more troubling. The good guys are the bad guys. Teaching your children to do the right thing can get people killed. We have other reasons for thinking the traditional model of how human beings work in ideal conditions (intellectual consideration of options informed by philosophical principles leading to rational action) may be not just flawed but downright wrong. Rai/Fiske suggest that the model is not even sustainable as a working hypothesis or *faute de mieux* but is downright dangerous.

INTERDISCIPLINARY SOCIAL RESEARCH

ZIYAD MARAR

Global publishing director, SAGE; author, *Intimacy*

In terms of sheer unfulfilled promise, interdisciplinary research has to stand as one of the most frustrating examples in the world of social research. The challenges modern society faces—climate change, antimicrobial resistance, countless issues to do with economic, social, political, and cultural well-being—do not come in disciplinary packages. They are complex and require an integrated response, drawing on different levels of inquiry. Yet we persist in organizing ourselves in academic siloes and risk looking like those blind men groping an elephant. As Garry Brewer pithily observed back in 1999, "The world has problems, universities have departments."

The reasons this promise is unfulfilled are equally clear. Building an academic career requires immersion in a speciality, with outputs (articles, books, talks) that win the approval of peers. Universities are structured in terms of departments, learned societies champion a single discipline, and funding agencies prioritize specific work from those who have built the right kind of credibility in this context. And this means interdisciplinary work is hard to do well, often falling between stools and sometimes lost in arcane debate about its very nature, swapping "inter" for "multi," "cross," "trans," "post," and other candidate angels to place on the head of this pin.

Some disciplines have overcome these hurdles—neuroscience,

bioinformatics, cybernetics, biomedical engineering—and more recently we have seen economics taking a behavioral turn and moral philosophy drawing on experimental psychology. But the bulk of the social sciences have proved resistant, despite the suitability of their problems to multilevel inquiry.

The good news is that we are seeing substantial shifts in this terrain, triggered in part by the rise of Big Data and new technology. Social researchers are agog at the chance to listen to millions of voices, observe billions of interactions, and analyze patterns at a scale never seen before. But to seriously engage requires new methods and forms of collaboration, with a consequent erosion of the once insurmountable barrier between quantitative and qualitative research. An example comes from Berkeley, where Nick Adams and his team are analyzing how violence breaks out in protest movements—an old sociological question, but now with a database (thanks to the number of Occupy movements in the U.S.) so large that the only way to analyze it feasibly requires a Crowd Content Analysis Assembly Line (combining crowd sourcing and active machine learning) to code vast corpora of text. This new form of social research, drawing on computational linguistics and computer science to convert large amounts of text into rich data, could lead to insights in a vast array of social and cultural themes.

These shifts might stick if we continue to see centers of excellence focusing on data-intensive social research, like D Base at Berkeley or the Harvard Institute for Quantitative Social Science, show how institutions can reconfigure themselves to respond to opportunity. As Gary King (director of the latter) has put it:

> The social sciences are undergoing a dramatic transformation from studying problems to solving them; from making do

with a small number of sparse data sets to analyzing increasing quantities of diverse, highly informative data; from isolated scholars toiling away on their own to larger scale, collaborative, interdisciplinary, lab-style research teams; and from a purely academic pursuit focused inward to having a major impact on public policy, commerce and industry, other academic fields, and some of the major problems that affect individuals and societies.

More structural change will follow these innovations. Universities around the world, having long invested in social-science infrastructure, are looking to these models. And we are seeing changes in funders' priorities, too. The Wellcome Trust, for instance, now offers the Hub Award to support work that "explores what happens when medicine and health intersect with the arts, humanities, and social sciences."

Of course the biggest shaper of future research is at the national level. In the U.K., the proposed implementation of a "cross-disciplinary fund" alongside a new budget to tackle "global challenges" may indicate the Government's seriousness of interdisciplinary intent. Details will follow, and they may prove devilish. But the groundswell of interest, sustained by opportunities in data-intensive research, is undeniable.

So interdisciplinary social research should increasingly become the norm, although specialization will still be important—after all, we need good disciplines to do good synthetic work. But we may soon see social sciences coalesce into a more singular social science and become more fully engaged with problem domains first and departmental siloes second.

INTELLECTUAL CONVERGENCE

ADAM ALTER

Psychologist; associate professor of marketing, Stern School of
Business, NYU; author, *Drunk Tank Pink*

Suppose a team of researchers discovers that people who earn
$50,000 a year are happier than people who earn $30,000 a year.
How might the team explain this result?

The answer depends largely on whether the team adopts a
telephoto zoom lens or a wide-angle lens. A telephoto zoom
lens focuses on narrower causes, like the tendency for financial stability to diminish stress hormones and improve brain
functioning. A team that uses this lens will tend to focus on
specific people who annually earn more or less money and any
differences in how their brains function and how they behave.
A team that adopts a wide-angle lens will focus on broader
differences. Perhaps people who earn more also live in safer
neighborhoods with superior infrastructure and social support.
Though each team adopts a different level of analysis and arrives at a different answer, both answers can be right.

For decades and even centuries, this is largely how the social
sciences have operated. Neuroscientists and psychologists have
peered at individuals through zoom lenses, while economists
and sociologists have peered at populations through wide-angle
lenses.

The big news of late is that these intellectual barriers are
dissolving. Scientists from different disciplines are either sharing
their lenses or working separately on the same questions and

then coming together to share what they've learned. Not only is interdisciplinary collaboration on the rise, but papers with authors from different disciplines are more likely to be cited by other researchers. The benefits are obvious. As the income-gap example shows, interdisciplinary teams are more likely to answer the whole question, rather than focusing on just one aspect at a time. Instead of saying that people who earn more are happier because their brains work differently, an interdisciplinary team is more likely to compare the roles of multiple causes in formulating its conclusion.

At the same time, researchers within disciplines are adopting new lenses. Social and cognitive psychologists, for example, have historically explored human behavior in the lab. They still do, but many prominent papers published in 2015 also included brain-imaging data (a telephoto zoom lens) and data from social-media sites and large-scale economic panels (wide-angle lenses). One paper captured every word spoken within earshot of a child during the first three years of his life, to examine how babies come to speak some words earlier than others. A second paper showed that research-grant agencies favor male over female scientists by examining the content of thousands of grant reviews. And a third analyzed the content of 47,000 tweets to quantify expressions of happiness and sadness. Each of these methods is a radical departure from traditional lab experiments, and each approaches the focal problem from an unusually broad or narrow perspective. These papers are more compelling because they present a broader solution to the problem they're investigating—and they're already tremendously influential, in part because they borrow across disciplines.

One major driver of intellectual convergence is the rise of Big Data, not just in the quantity of data but also in understanding how to use it. Psychologists and other lab researchers have begun to complement lab studies with huge, wide-angle social-media and panel-data analyses. Meanwhile, researchers who typically adopt a wide-angle lens have begun to complement their Big Data analyses with zoomed-in physiological measures, like eye-tracking and brain-imaging analyses. The news here is not just that scientists are borrowing from other disciplines but also that their borrowing has supplied richer, broader answers to a growing range of important scientific questions.

WEAPONS TECHNOLOGY POWERED HUMAN EVOLUTION

TIMOTHY TAYLOR

Professor of the prehistory of humanity, University of Vienna; author, *The Artificial Ape*

Thomas Hobbes's uncomfortable view of human nature looks remarkably prescient in the light of new discoveries in Kenya. Back in the mid-17th century—before anyone had any inkling of deep time or the destabilization of essential identity that would result from an understanding of the facts of human evolution (i.e., before the idea that nature was mutable)—Hobbes argued that we were fundamentally beastly (selfish, greedy, cruel) and in the absence of certain historically developed and carefully nurtured institutional structures, we would regress to live in a state of nature, in turn understood as a state of perpetual war.

We can assume that John Frere would have agreed with Hobbes. Frere, we may read (and here *Wikipedia* is orthodox, typical, and, in a critical sense, wrong), "was an English antiquary and a pioneering discoverer of Old Stone Age or Lower Paleolithic tools in association with large extinct animals at Hoxne, Suffolk in 1797." In fact, while Frere did indeed make the first well-justified claim for a deep-time dimension to what he carefully recorded *in situ*, saying that the worked flints he found dated to a "very remote period indeed," he did not think they were tools in any neutral sense, stating that the objects were "evidently weapons of war, fabricated and used by a people who had not the use of metals."

Frere's sharp-edged weapons can now be dated to Oxygen Isotope Stage 11—that is, to a period lying between 427,000 and 364,000 years ago—and even he might have been surprised to learn that the people responsible were not modern humans but a species, *Homo erectus*, whose transitional anatomy would first come to light through fossil discoveries in Java a century later. Subsequent archeological and paleoanthropological work (significant aspects of it pioneered by Frere's direct descendant, Mary Leakey) has pushed the story of genus *Homo* back ever further, revealing as many as a dozen distinct species (the number varies with the criteria used).

Alongside the biological changes runs a history, or prehistory, of technology. It has usually been supposed that this technology, surviving mainly as modified stone artifacts, was a product of the higher brain power that our human ancestors displayed. According to Darwin's sexual-selection hypothesis, female hominins favored innovative male hunters, and the incremental growth in intelligence led, ultimately, to material innovation (hence the evocative but not taxonomic term *Homo faber*—Man the maker).

So powerful was this idea that although chipped-stone artifacts dating to around 2.6 million years ago have long been known, there was a strong presumption that genus *Homo* had to be involved. This is despite the fact that the earliest fossils with brains big enough to be classified in this genus dated at least half a million years later. It was a fairly general hunch within paleoanthropology that the genus *Homo* populations responsible for early chipped-stone technologies had simply not yet been discovered. Those few of us who, grounded in the related field of theoretical archeology, thought differently remained reliant on a broad kind of consilience to counter this *ex-silencio* assumption.

So to me it was wonderful news when in May 2015 Sonia Harmand and co-workers published an article in *Nature* titled "3.3-million-year-old stone tools from Lomekwi 3, West Turkana, Kenya"—because the strata at their site date to a period when no one seriously doubts that australopithecines, with their chimp-sized brains, were the smartest of the savannah-dwelling hominins. The discovery shows unambiguously that technology preceded, by more than a million years, the expansion of the cranium traditionally associated with the emergence of genus *Homo*.

Harmand *et al.* follow current convention in naming their artifacts "tools" rather than weapons. This more neutral terminology is, I think, a reflex of our continual and pervasive denial of Hobbes's truth. Hobbes obviously did not have the information available to understand the time-depths or the bio-taxonomic and evolutionary issues involved, but had he known about Lomekwi 3 he would probably have called it differently, as Frere subsequently did at Hoxne.

The prehuman emergence of weapons technology makes functional and evolutionary sense. We're not talking about choosing appropriate twigs for the efficient extraction of termites, or the leaves best suited to carrying water, but about the painstaking modification of fine-grained igneous rocks into sharp-edged or bladed forms whose principal job was to part flesh from flesh and bone from bone.

If the word "hunting" helps gloss over the reality of what such sharp stone artifacts were used for, then we can reflect on the type of hunting recently documented from the Aurora stratum at the site of Gran Dolina (Atapuerca), Spain. Here, Palmira Saladie Balleste and co-workers recently reported (*Journal of Human Evolution*, November 2012) on a group of *Homo anteces-*

sor (or, arguably, *erectus*) who fed on individuals from, presumably, a rival group. The age profile of the victims was "similar to the age profiles seen in cannibalism associated with intergroup aggression in chimpanzees"—that is, those eaten were infants and immature individuals.

That much of the data on early systematic endemic violence in our deeper (and shallower) prehistory comes as a surprise is due to the pervasive myth about the small-scale band societies of the archeological past having somehow been egalitarian. This idea has an odd genealogy that probably goes back to the utopian dreaming of Hobbes's would-be intellectual nemesis, Jean-Jacques Rousseau, with his ideas of harmony and purity in nature and his belief that the savage state of humanity was noble rather than decadent.

Rousseau, like Hobbes, lacked an evolutionary perspective and also dealt in more or less essentialist assertions. But once we have to bridge from wild primate groups to early human ancestors, the tricky question of the roots of egality arises. If the chimpanzees and gorillas we study in the wild have clear status hierarchies that are established, maintained, and altered by force, including orchestrated murder and cannibalism, then how could fair play magically become a base-line behavior?

Somewhere along the way, we seem to have forgotten that the 19th-century founders of modern sociocultural anthropology, Lewis Henry Morgan, Edward Burnett Tylor, and Edvard Westermarck, recorded all kinds of unfairness in indigenous North American and Pacific societies. We may not want the Tlingit and Haida, Ojibwa and Shawnee to have had slaves, for instance, but avoiding mentioning it in modern textbooks does a great disservice to the original ethnographic accounts (and the slaves).

Returning to the news from Lomekwi 3, we can now see that technology was not just a figurative but a literal arms race. Re-analyzing the increase in cranial capacity that began around 2 million years ago, we can see that blades and choppers needed to be already available to replace the missing biology—the massive ripping canines and heavy jaw muscles that previously hampered, in bio-mechanical terms, the expansion of the braincase. As Frere would immediately have understood, these artifacts included weapons, perhaps predominantly. Only by postulating high-level competition between groups can we understand the dramatic adaptive radiation of hominin types and the fact that, ultimately, only one hominin species survived.

The paradox is that by sharpening the first knives to extend the range of possible forms of aggression, we opened up a much broader horizon, in which technology could be used for undreamed-of purposes. Yet Hobbes's instinct that our nature was borne of war, and Frere's conclusion that the world's primal technology was offensive, should not be ignored. In the artificial lulls when atavism is forced into abeyance, we are happy to forget Hobbes's admonition that it is only through the careful cultivation of institutions that stable peace is at least possible.

THE IMMUNE SYSTEM: A GRAND UNIFYING THEORY FOR BIOMEDICAL RESEARCH

BUDDHINI SAMARASINGHE

Molecular biologist; science communicator; co-creator of the Web site *Know the Cosmos*

The germ theory of disease launched a revolution that transformed medicine. For the first time in history, disease was understood as an attack by microscopic organisms, organisms that were soon identified, characterized, and defeated. Yet we have not conquered disease. Cancer, heart disease, diabetes, stroke, and parasitic diseases are some of the top causes of death in the world today. Is there a unifying principle explaining them all? A common mechanism that could help us transform biomedical science once again? Perhaps there is: the immune system.

We are just beginning to understand how deeply involved the immune system is in our lives. Its cellular sentries weave an intricate early warning network throughout the body; its signaling molecules—the cytokines—trigger and modulate our response to infection, including inflammation; it is involved even in as humble a process as the clotting of blood in a wound. At a fundamental level, the immune system provides us with a framework to better understand and treat a wide range of ailments.

Cancer is often described as a "wound that never heals," referring to the chronic inflammatory state that promotes tumor development. Cancer cells develop into a tumor by disabling and hijacking components of the immune system; immunosuppression and tumor-promoting inflammation are the two facets of cancer immunology. Both Type 1 and Type 2 diabetes are linked to the immune system; the former is an autoimmune condition in which the immune system attacks the insulin-producing cells in the pancreas; the latter is linked to insulin resistance through high levels of cytokines produced during inflammation. Pro-inflammatory cytokines are also linked to heart disease, which is the leading cause of death in the developed world. The malaria parasite is an expert at manipulating our immune system, cloaking itself in molecules that reassure our immune sentinels that nothing is amiss, while wreaking havoc in our blood cells.

Most intriguing, we are now beginning to learn about the neuroimmune system, a dense network of biochemical signals synthesized in neurons, glial cells, and immune cells in our brains, and critical to the function of our central nervous system. These markers of the neuroimmune system are disrupted in disorders such as depression, anxiety, stroke, Alzheimer's disease, Parkinson's disease, and multiple sclerosis. Cytokine levels have been shown to vastly increase during depressive episodes, and—in people with bipolar disorder—to drop off in periods of remission. Even the stress of social rejection or isolation causes inflammation, leading to the fascinating idea that depression could be viewed as a physiological allergic reaction rather than simply a psychological condition.

With this knowledge comes power: Modulating the immune system to our advantage is a burgeoning field of research, particularly for cancer. Cancer immunotherapy marks a turning point in treatment, with astonishingly rapid remissions achieved in some patients undergoing early-stage clinical trials. New classes of drugs known as checkpoint inhibitors target specific immunological pathways, and we can reprogram "designer" immune cells to target cancer cells. Aspirin, a humble drug that reduces inflammation, may even be able to prevent some cancers—a tantalizing possibility currently being investigated in a large-scale clinical trial. Aspirin is already known to prevent heart attack and stroke in some people, through its anti-inflammatory and anti-clotting effects. Fascinating studies imply that supplementing antidepressants with anti-inflammatory drugs can improve their efficacy. Vaccines for such infectious diseases as malaria and HIV-AIDS are imminent. These advances couldn't have come at a better time, since antibiotics have become increasingly ineffective due to widespread resistance (a problem classified by the World Health Organization as a "global threat"). Fields are converging in ways we haven't seen previously; oncologists, parasitologists, neurobiologists, and infectious-disease specialists are all collaborating with immunologists.

It is an exciting time in biology and medicine. The new discoveries about the breadth and potential of our immune response merely hint at revelations to come. These research findings will always be newsworthy, because they promise to help us endure beyond disease and enjoy a longer life. Although we should beware the sham medicine of "miracle cures" and "immune boosting diets," if we can drive research forward while communicating it effectively, we may be on the cusp of another revolution in biomedical science.

HARNESSING OUR NATURAL DEFENSES AGAINST CANCER

MICHAEL E. HOCHBERG

Population biologist, Centre National de la Recherche Scientifique,
University of Montpellier, France

One out of every two people will have to deal with a diagnosis
of cancer during their lifetime. The 10 percent of cancers that
arise in genetically high-risk groups alone represents less than
1 percent of the total U.S. population but costs a staggering $15
billion to treat annually.

Despite decades of research, the Holy Grail of a cure still
eludes us, in part because of the fundamental unstable nature of
cancer: It easily produces variant cells that resist chemotherapies, and this often results in relapse. Cancer is also difficult because cancer types differ considerably in their biology, meaning
that a single drug is unlikely to be effective against more than
one or a few types. Finally, even within a cancer type, patients
can have differences in how their cancers react to a given drug.
What all this means is that "one drug cures all" is not one drug
and unfortunately for metastatic cancer, often not a cure.

The problem can be stated thus, in brief: For late-stage cancers, which are the most difficult to treat, most new drugs are
considered a success if they extend life for several weeks or
months. The limited or disappointing results of many chemotherapies has led to concerted efforts to identify the Achilles'
heel, or rather heels, of cancers. To gauge where the most promising discoveries are being made, just read the titles any week

of the world's most prestigious scientific journals and parallel coverage in the popular press: It's all about immunotherapies.

The idea makes sense: Harness a patient's own natural mechanisms for eliminating diseased cells, or give the patient man-made immune-system components to help specifically target malignancies. This is certainly better, all else being equal, than injections of toxic drugs. The basic challenge of traditional chemotherapies is that they affect both cancerous and, to some extent, healthy cells, meaning that for the drugs to work, doses must be carefully established to kill or arrest the growth of cancer cells, while keeping the patient alive. The more drug, the more the cancer regresses, but the higher the chance of side effects or even patient death. Many patients cannot withstand the doses of chemotherapies most likely to cure them, and even if they can, exposing rapidly dividing, mutation-prone cancer cells will strongly select for resistance to the therapy—which is why remission is often followed by relapse.

Employing our own immune systems has intuitive appeal. Our bodies naturally use immuno-editing and immune surveillance to cull diseased cells. However, the tumor microenvironment is a complex adaptive structure that can also compromise natural and therapy-stimulated immune responses. This past year has seen important milestones. For example, based on promising clinical trials, the FDA recently approved a combination of two immunotherapies (Nivolumab and Ipilimumab) for metastatic melanoma. What one is not able to accomplish, the other is; this not only reduces tumor size but is expected to result in less evolved resistance to either drug. The same idea of using combinations can be applied to immunotherapies together with many of the more traditional radiotherapies, chemotherapies, and more recent advances in targeted therapies.

Currently more than forty clinical trials are being conducted to examine effects of immunotherapies on breast cancer, with the hope that within a decade such therapies, if promising, can reduce or eliminate these cancers in the one in eight women who currently are affected during their lifetimes. This would be truly amazing headline news.

CANCER DRUGS FOR BRAIN DISEASES

TODD C. SACKTOR

Distinguished Professor of physiology, pharmacology, and neurology,
SUNY Downstate Medical Center

There has not been a new effective therapy for any neurodegenerative disease in decades. Recent trials of drugs for Alzheimer's disease have been disappointments. Because of these expensive failures, many of the big pharmaceutical companies have moved away from targeting brain diseases to more profitable areas, like cancer. So is there any good news on the horizon for the millions who are suffering and will suffer from these devastating brain disorders?

In 2015, there was news that a cancer drug showed remarkable benefits for patients with Parkinson's disease. It was only one nonrandomized, nonblinded, non–placebo-controlled study that looked at only a few patients, so it's too early to know whether it really works. But this is news to follow, and it's big news for three reasons.

First, unlike any other treatment, the drug appears to work close to the root cause of Parkinson's. In Parkinson's, the neurons that supply the brain with the neurotransmitter dopamine degenerate. The mainstay of treatment for the disease—Parkinson's is one of the few neurodegenerative disorders for which there is any effective treatment—has been to replace that missing dopamine with a pill that provides a chemical that converts to dopamine in the brain. This therapy treats the symptoms of Parkinson's—the tremors, the stiffness, and the slowness of movements—but not its root cause. So the death of the dopamine-containing neurons continues unabated and the pills work well only for around seven years.

The new drug, called nilotinib, was developed for leukemia and has the same action as the better-known chemotherapeutic agent Gleevec. But unlike other similar drugs, nilotinib gets across the blood-brain barrier, which prevents most drugs from working well in the brain. Although the cause of the neuronal degeneration of Parkinson's is still unknown, it is thought to involve the accumulation and misfolding of proteins inside the dying neurons, a process like the curdling of the proteins in milk. Nilotinib was predicted to suppress the accumulation of misfolded proteins inside neurons. After taking nilotinib, the patients not only did better clinically but the amount of the misfolding proteins released into their cerebral spinal fluid went down—a sign that it was working on the degenerative process itself.

Second, the target that nilotinib inhibits is a new one for a brain disease. Like Gleevec, nilotinib inhibits an enzyme inside the cell called a protein kinase. There are around 500 different kinds of protein kinases in cells, and nilotinib targets one of them. Whereas there are many kinases in a cell, there are far more biochemical functions a cell has to do. So most kinases have multiple functions, some seemingly unrelated. Scientists focused on the kinase that nilotinib inhibits because if it becomes overactive, it can drive unchecked growth of white cells in the blood, causing leukemia. But they also found that it's involved in the accumulation of neuronal proteins that can get misfolded. Nilotinib is big news because drugs that target kinases are relatively easy to develop, and nilotinib provides the first example showing that if they work for one disease, they might be used for a second seemingly unrelated disease. At the bedside, leukemia and Parkinson's seem as far apart as you can get.

Third, the timing with which the drug may work tells us something new and exciting about Parkinson's itself, which might be relevant to other neurodegenerative diseases, such as Alzheimer's. Protein misfolding in neurons is a general process in many neurodegenerative disorders. But no one knows whether suppressing protein misfolding will result in the slowing or stopping of a disease, or even in recovering function. The effect of nilotinib seems relatively fast—the trial lasted only a few months. If nilotinib's beneficial action is really on inhibiting the accumulation and misfolding of neuronal proteins (and not secondarily, on increasing the release of dopamine), and if the patients improved, this could mean that the misfolding is one side of an active and dynamic battle in neurons between "good" folding and "bad" folding. In that case, we would conclude that there are neuronal processes that are actively trying to repair the cell. This gives us hope for a cure and restoration of lost function in many neurological diseases.

THE MOST POWERFUL CARCINOGEN MAY BE ENTROPY

GEORGE JOHNSON

Science writer; columnist, *New York Times*; author, *The Cancer Chronicles*

Cancer is often described as a sped-up version of Darwinian evolution. Through a series of advantageous mutations, the tumor—this hopeful monster—becomes fitter and fitter within the ecosystem of your body. Some of the mutations are inherited, while others are environmental—the result of a confusion of outside influences. Much less talked about is a third category: the mutations that arise spontaneously from the random copying errors occurring every time a cell divides.

In a recent paper in *Science*, Cristian Tomasetti and Bert Vogelstein calculated that two-thirds of the overall risk of cancer may come from these errors—entropic "bad luck."* The paper set off a storm of outrage among environmentalists and public health officials, many of whom seem to have misunderstood the work or deliberately misrepresented it. And a rival model has since been published in *Nature* claiming to show that, to the contrary, as much as 90 percent of cancer is environmentally caused.† That to me is the least plausible of these dueling reports. As epidemiology marches on, the link between cancer

---

* http://www.sciencemag.org/content/347/6217/78.abstract

† http://www.nature.com/nature/journal/v529/n7584/full/nature16166.html

and carcinogen grows ever fuzzier. The powerful and unambiguous link between smoking and lung cancer seems almost a fluke.

It will be interesting to see how this plays out. But meanwhile I hope that more of the public is beginning to understand that getting cancer usually doesn't mean you did something wrong or that something bad was done to you. Some cancer can be prevented and some can be successfully treated. But for multicellular creatures living in an entropic world, a threshold amount of cancer is probably inevitable.

THE DECLINE OF CANCER

A. C. GRAYLING

Philosopher; master, New College of the Humanities, and supernumerary fellow, St. Anne's College, Oxford, U.K.; author, *The God Argument*

In a great year for science, it is hard to restrict applause to just one area, but a worthy one is cancer research, which has seen a number of advances. Genetic manipulation has rapidly reversed colorectal cancer in mice, a Dutch team has developed a highly accurate blood test for cancer, a general cure for cancer is promised by the discovery that attaching malaria proteins to cancerous cells destroys them, the Mayo Clinic has found a way of short-circuiting cancer-cell growth by using a certain junction protein (PLEKHA7), early diagnosis of pancreatic cancer looks more possible following the identification of a protein on particles released by cancer cells in the pancreas, low-toxicity nanopills for treatment of breast cancer look to be on the way, and the FDA has approved Palbociclib for breast cancer treatment. There may have been other announcements in the oncology field this year, but the cumulative effect of these developments would appear to support the claim made by a leading oncologist that within a generation no one under the age of eighty will die of cancer.

THE MATING CRISIS AMONG EDUCATED WOMEN

DAVID M. BUSS

Professor of psychology, University of Texas, Austin; co-author (with
C. M. Meston), *Why Women Have Sex*

Every year, more women than men become college-educated.
The disparity is already prevalent in North America and Europe,
and the trend is beginning to spread across the world. At the
University of Texas at Austin, where I teach, the sex ratio is
54-percent women to 46-percent men. This imbalance may not
seem large at first blush. But when you do the math, it translates
into a hefty 17-percent more women than men in the local
mating pool. Speculations about reasons range widely. They include the gradual removal of gender discrimination barriers and
women's higher levels of conscientiousness (relative to men's),
which translates into better grades and superior college-app
qualifications. Whatever the causes, the disparity is creating a
mating crisis among educated women.

We must look deeply into our mating psychology to understand the far-reaching consequences of the sex-ratio imbalance.
Women and men both have evolved multiple mating strategies. Some of each gender pursue casual hook-ups; some pursue
committed partnerships. Some alternate at different times of
their lives; some do both simultaneously. And although a few
social scientists deny the data, research overwhelmingly shows
that men harbor, on average, a greater desire for sexual-partner
variety. Men experience more frequent sexual thoughts per

day, have more sexual fantasies involving multiple partners, and more readily sign up for online dating sites for the sole goal of casual sex. Thus a surplus of women among educated groups caters precisely to this aspect of men's sexual desires, because the rarer gender is always better positioned to get what it wants on the mating market. In places like large cities in China, with their surplus of men, women can better fulfill their desires while many men remain frustrated and mateless. Context matters. For every surplus of women in places like Manhattan, there exist pockets where men outnumber women, such as schools of engineering or the software companies of Silicon Valley. But when there are not enough men to go around, women predictably intensify their sexual competition. The rise of hook-up cultures on college campuses and online dating sites like Tinder, Adult Friend Finder, and Ashley Madison is no coincidence.

Gender differences in sexual psychology are only part of the problem. Additional elements of the mating mind exacerbate it. A key cause stems from the qualities women seek in committed mateships. Most women are unwilling to settle for men who are less educated, less intelligent, and less professionally successful than they are. The flip side is that men are less exacting about precisely those criteria, choosing to prioritize, for better or worse, other factors, such as youth and appearance. So the initial sex-ratio imbalance among educated groups gets worse for high-achieving women. They end up being forced to compete for the limited pool of educated men not just with their more numerous educated rivals but also with less-educated women, whom men find desirable for other reasons.

The relative shortage of educated men worsens when we add the factors of age and divorce to the mating matrix. As men age, they desire women increasingly younger than they are. Intelligent, educated women may go for a less accomplished partner for a casual fling, but for a committed partner they typically want mates their own age or a few years older and at least as educated and career-driven. Since education takes time, the sex-ratio imbalance is especially skewed among the highly educated—those who seek advanced degrees to become doctors, lawyers, or professors, or who climb the corporate ladder post-MBA. And because men are more likely than women to remarry following divorce and to marry women increasingly younger than they are—three years at first marriage, five at second, eight at third—the gender-biased mating ratio skews more sharply with increasing age.

Different women react in different ways to the mating crisis. Some use sexual tactics to ramp up their competition for men. They dress more provocatively, send more sexually explicit texts, consent to sex sooner, and hope that things turn into something more than a brief encounter. Some women opt out of the mating game, unwilling to compromise their careers in the service of mating. (Although some progress has been made, women still suffer disproportionately from compromises between career and family.) And some women hold out for an ever smaller pool of men who are single, educated, and emotionally stable; who are not sexual players; and who can engage their intellect, sense of humor, emotional complexities, and sexual passions for more than just a night.

The good news for those who succeed is that marriages among the educated tend to be more stable, freer of conflict, less plagued by infidelity, and less likely to end in divorce. Educated couples enjoy a higher standard of living, as dual professional incomes catapult them to the more affluent tiers of the economy. They suffer less financial stress than their less educated counterparts. Assortative coupling on the education level does have an unintended downside—it's a major contributor to economic inequality in the larger society, widening the gap between the haves and have-nots. But for accomplished women who successfully overcome the odds unfairly stacked against them, mating triumph typically takes precedence over loftier goals of reducing societal-level inequality when the two conflict.

What are the potential solutions to the mating-pool shortage for educated women? Should they adjust their mate preferences? Expand the range of men they're willing to consider as mates? Mating psychology may not be that malleable. The same mating desires responsible for the skewed gender imbalance to begin with continue to create unfortunate obstacles to human happiness. As successful women overcome barriers in the workplace, they encounter new dilemmas in the mating market.

THE MOST IMPORTANT
X . . . Y . . . Z . . .

JARED DIAMOND

Professor of geography, UCLA; author, *The World Until Yesterday*

In many fields, one hears questions in the format, "What is *the* most important X . . . Y . . . Z . . . , etc.?" For instance, what is *the* most important factor accounting for artistic creativity? Or competitive biological success? Or a happy marriage? Or military success? Or scientific creativity? Or successful child-rearing? Or a sustainable economy? Or world peace?

In our complicated, multifactorial world, the correct answer to such a question is almost always, "The most important consideration is not to search for *the* most important consideration." Instead, there are normally many considerations, none of which can be ignored.

For instance, marital therapists have identified nineteen independent factors essential to a happy marriage: compatibility about sex, money, religion, politics, in-laws, child-rearing, styles of arguing, and twelve other factors. If a couple agrees about eighteen of those factors but can't resolve a disagreement just about sex (or just about money, or just religion, etc.), they are in deep trouble. Hence if you hear a newly married couple ask you in all seriousness, "What is the single most important requirement for a happy marriage?" you can bet that that marriage will end in divorce.

THE MOTHER OF ALL ADDICTIONS

HELEN FISHER

Biological anthropologist, Rutgers University; senior research fellow, the Kinsey Institute; author, *Why Him? Why Her? How to Find and Keep Lasting Love*

Falling in love activates the same basic brain system for wanting (the reward system—specifically the mesolimbic dopamine pathway), as do all drugs of abuse, including heroin, cocaine, alcohol, and nicotine. Because this central neural network becomes active when addicted to *any* drug of abuse, I have long wondered whether feelings of romantic love can smother a drug craving or whether a drug craving can smother feelings of romantic love, or whether these two very different cravings might work together—sensitizing this brain network to make the drug addict more receptive to romantic love and/or make the lover more prone to other forms of addiction. In short: How does this central brain system accommodate two different cravings at once?

All these questions are still largely unanswered. But in 2012 an article made inroads into this conundrum. Xiaomeng Xu and her colleagues put eighteen Chinese nicotine-deprived smokers who had also just fallen madly in love into a brain scanner, using functional magnetic resonance imaging (fMRI).* As these men and women looked at a photo of a hand holding a cigarette and

---

* http://journals.plos.org/plosone/article?id=10.1371/journal.pone.0042235

also at a photo of their newly beloved, the researchers collected data on their brain activity. Results? Among those who were moderately addicted to nicotine, the craving for the beloved reduced activity in brain regions associated with the craving for a cigarette.

But there is some added value here. The article also suggests that engaging in any kind of novel activity (unrelated to romance) may also alleviate nicotine craving—by hijacking this same dopaminergic reward system. This single correlation could be of tremendous value to those trying to quit smoking.

And I shall go out on a limb to propose a wider meaning to these data. Although there is only this very limited evidence for my hypothesis, this study suggests to me that there may be a hierarchy to the addictions. In this case, one's addiction to a newly beloved may, in some cases, suppress one's addiction to nicotine. Romantic love may be the mother of all addictions—indeed, a *positive* addiction enabling one to overcome other cravings to win life's greatest prize: a mating partner.

THE TRUST METRIC

JOHN GOTTMAN

Psychologist; cofounder, The Gottman Institute; author, *The Seven Principles for Making Marriage Work*

What has amazed and excited me the most in recent scientific news is that the concept of trust can be measured validly and reliably and that it organizes a vast amount of information about what makes families and human societies function well, or fail.

As a relationship researcher and couples/family therapist, I have known for decades that trust is the number-one issue that concerns couples today. Consistent with this is the finding that the major trait people search for in trying to find a mate is trustworthiness. Robert Putnam's groundbreaking book *Bowling Alone* began documenting this field of scientific research, which is based on a very simple question. Sociologists have used a yes/no survey question: "In general, would you say that you trust people?" It turns out that regions of the U.S., and countries throughout the world, vary widely in the percentage of people who answer "Yes."

Here's the amazing scientific news. In regions of the U.S., the percentage of people who trust others correlates highly with an array of positive social indices such as greater economic growth, greater longevity of citizens, their better physical health, lower crime rates, greater voting participation, greater community involvement, more philanthropy, and higher student achievement scores—to mention just a few variables that index the health of a community. As we move from the North to the South in the United States, the proportion of people who trust others drops continuously. A great archival index of trust turns out to be the discrepancy in income between the richest and poorest people in a region.

High income discrepancy implies low trust. That discrepancy has been growing in the U.S. since the 1950s, as has the decline in community participation. Data show that in the 1950s CEOs earned about 25 times more than the average worker; that ratio has grown steadily, so that in 2010 it was about 350 times more. So we are in a crisis in this country, and it's no surprise that the gap between the rich and the poor has become a major issue in the 2016 election. One fact in these results: How well the U.S. cares for its poorest citizens is a reliable index of the social and economic health of the entire country. Empathy for the poor is thus smart politics.

These results also hold internationally, where the trust percentage is also related to less political corruption. Only 2 percent of the people in Brazil trust one another, whereas in Norway 65 percent trust others. Many other factors are important internationally, but we note that Brazil is currently in chaos, while Norway is thriving.

These spectacular data are, unfortunately, correlational. Of course it's hard to do real experiments at societal levels; however, these findings on trust have spawned growing academic fields of behavioral economics and neuro-economics, which fields are generating exciting new experiments. Combined with the mathematics of game theory, this work has led to the creation of a valid "trust metric" in interactions between two people. A new understanding of the processes of how two people build (or erode) trust in a love relationship has enabled a new therapy currently being tested.

We are approaching an understanding of human cooperation in family relationships that generalizes to society as a whole. I'm hopeful that these breakthroughs may eventually lead us to form a science of human peace and harmony.

OPTOGENETICS

CHRISTIAN KEYSERS

Neuroscientist; director, Social Brain Lab, Netherlands Institute for
Neuroscience; author, *The Empathic Brain*

Over the past decade, with the discovery of optogenetics, neuroscience has thrown open a door that seemed closed forever.
Before optogenetics, our ability to record the activity of cells in
the brain was sophisticated, and we understood that our emotions, our thoughts, and our perceptions entail the activity of
millions of cells. What we lacked was the ability to trigger a
similar state in the brain on command. Neuroscience was a spectator of the mind, not an actor. With the advent of optogenetics,
this is changing.

Optogenetics is a new field of biotechnology that lets us
transform brain activity into light and light into brain activity.
It allows us to introduce fluorescent proteins into brain cells
to make them glow when they're active, thereby transforming
neural activity into light. It also allows us to introduce pho-
tosensitive ion channels into neurons, so that shining light on
the cells triggers activity or silences neurons at will—thereby
transforming light into neural activity. By employing modern
technologies that record light from neurons deeper and deeper
in the brain and guide light onto individual neurons, we have
crossed a frontier that only a decade ago seemed far away: For
the first time, we can selectively re-create arbitrary states in the
brain—and hence the mind.

A small number of experiments have demonstrated the potential of this technique. For instance, mice were made to experience fear, and then optogenetics reactivated the pattern of neural activity triggered during the original experience and the mice froze in fear again. Neuroscience has become a protagonist. The science-fiction scenario of "total recall" (in which Arnold Schwarzenegger was implanted with memories he never had, in the film of the same name) now becomes practicable. In another set of experiments, the activity of cells in one animal's brain was recorded and imposed on corresponding cells in the brain of another animal—which was then able to take decisions based on what the other animal was feeling.

I predict that the ability to measure and re-create brain activity at the level of specific neurons is about to transform us in ways no other invention has. The invention of fire, of the wheel, of antibiotics, of the Internet, changed our lives in profound ways, making them safer, more comfortable, more exciting. But they have not changed who we are. Recording and manipulating brain activity will change who we are. It will serve as an interface through which computers can become part of our brain and through which our brains can directly interface with each other.

When we observe a baby grow into a child, we see how profoundly a person changes when connections in her brain allow her to tap into the resources of new brain regions. Soon, for some of us, this process will continue beyond the confines of our body, when optogenetic-like technologies allow our minds to encompass the world of computers. Who will we become? What will the world look like, sensed directly not only with our own senses but with all the sensors of the Internet of Things? What would global climate negotiations be like if we could directly connect with the brains of the people around us in the ultimate form of empathy? How will our societies deal with a transition phase in which neuro-enhancement will be affordable to some of us and not to others? In which some will have amazing powers of thought while others remain confined within their own brains?

THE STATE OF BRAIN SCIENCE

TERRENCE J. SEJNOWSKI
Computational neuroscientist; Francis Crick Professor, Salk Institute;
co-author (with Patricia Churchland), *The Computational Brain*

The big news on April 2, 2013, was the announcement of the BRAIN Initiative from the White House; its goal is to develop innovative neurotechnology for understanding brain function. Grand challenges like this happen once every few decades, including the announcement in 1961 of the Apollo Program to land a man on the Moon, the War on Cancer in 1971, and the Human Genome Project in 1990. These were ten-to-fifteen-year national efforts bringing together the best and the brightest to attack a problem that could be solved only on a national scale.

Why the brain? Brains are the most complex devices in the known universe and so far our attempts to understand how they work have fallen short. It will take a major international effort to crack the neural code. Europe weighed in earlier with the Human Brain Project and Japan later announced a Brain/MINDS project to develop a transgenic nonhuman primate model. China is also planning an ambitious brain project.

Brain disorders are common and devastating. Autism, schizophrenia, and depression destroy personal lives and place an enormous economic burden on society. In this country, the annual cost of maintaining patients with Alzheimer's disease is $200 billion and climbing as our population ages. Unlike heart diseases and cancers that lead to rapid death, patients can live for decades with brain disorders. The best efforts of drug companies to develop new treatments have failed. If we don't find a better way to treat broken brains now, our children will suffer terrible economic consequences.

A second reason to reach a better understanding of brains is to avert a catastrophic collapse of civilization, which is happening in the Middle East. The Internet has enabled terrorist groups to proliferate, and modern science has created existential threats ranging from nuclear weapons to genetic recombination. The most versatile weapon-delivery system is the human being. We need to better understand what happens in the brain of a suicidal terrorist planning maximum destruction.

These motivations are based on brains behaving badly, but the ultimate scientific goal is to discover the basic principles of normal brain function. Richard Feynman once noted that "what I cannot create, I do not understand." That is, if you can't prove something to yourself, you don't really understand it. One way to prove something is to build a device based on what you understand and see if it works. Once we have truly uncovered the principles of how the brain works, we should be able to build devices with similar capabilities. This will have a profound effect on every aspect of society, and the rise of artificial intelligence based on machine learning is a harbinger. Our brain is the paramount learning machine.

These are the goals of the BRAIN Initiative, but its results may be different from our expectations. The goal of the Apollo Program was accomplished, but if the Moon was so important why have we not gone back there? In contrast, building the technologies needed to reach the Moon has produced many benefits: a thriving satellite industry; advances in digital communications, microelectronics, and materials science; and a revamping of the science and engineering curriculum.

The War on Cancer is still being fought, but the invention of recombinant DNA technology allowed us to manipulate the genome and created the biotechnology industry. The goal of the Human Genome Project was to cure human diseases, which we now know are not easily deciphered by reading the base pairs, but the sequencing of the human genome has transformed biology and created a genomic industry that fosters personalized, precision medicine.

The impact of the BRAIN Initiative will be the creation of neurotechnologies that match the complexity of the brain. Genetic studies have uncovered hundreds of genes that contribute to brain disorders. Drugs have not been as effective in treating brain disorders as they have for heart diseases, because of the diversity of cell types in the brain and the complexity of the signaling pathways. The development of new neurotechnologies will create tools that can more precisely target the sources of brain disorders. Tools from molecular genetics and optogenetics are already giving us an unprecedented ability to manipulate neurons, and more powerful tools are on the way from the BRAIN Initiative.

An important lesson from the history of national grand challenges is that there is no better way to invest in the future than focusing the best and brightest minds on an important problem and building the infrastructure needed to solve it.

NOOTROPIC NEURAL NEWS

GEORGE CHURCH

Professor of genetics, Harvard Medical School; director, Personal
Genome Project; co-author (with Ed Regis), *Regenesis*

The most accessed parts of the Internet focus on new news
and old news via search engines and social-network news about
shopping, pets, and humans—especially sportful and celebrity
humans. What is the distinction between popularity and enduring importance?

In remote indigenous peoples (300 million strong, including Kawahiva, Angu, Sentineli) and our primate relatives, the
distinction seems small. In contrast, in our hypercivilization,
the importance of survival has been decoupled from popularity. Our ancient starvation for sugar and fat has morphed today
into nearly limitless *ad libitum* cardio-challenging doughnuts and
steaks. Our instincts to reproduce can now be rechanneled into a
wide variety of diversions. Practice for the hunt with rocks and
spears has become inflated to fill 514 stadiums holding 40,000 to
220,000 spectators, with up to 4.8 billion viewers via electronics.
Mild analgesic herbal medicines have become powerfully pure
and addictive. Running toward (or away from) a predator-prey
encounter has transformed into a market for fast cars, killing
1.2 million people per year (roughly equal to all humans alive
10,000 years ago).

Our Darwinian drive to improve our survival relative to
other species now includes augmentations that would be baffling to our ancestors—dodging asteroids via Mars colonies

and handheld neural prosthetic supercomputers with two video cameras.

The new news is that Greenpeace, KMP, and MASIPAG are accused of "crimes against humanity" for blocking (including vandalizing safety-testing experiments), from 2002 to 2016, golden rice, which could save a million souls per year from vitamin-A deficiency.

The old news, again (courtesy of the national academies of the U.S., U.K., and China), is that after forty years we still haven't reached a consensus on whether we want embryo (germline) augmentation. But this is likely a moot point, since genetic and non-genetic adult augmentation represents hundredfold larger markets and much faster potential return on enhancement—weeks rather than decades; Web-warp-drive speed vs. human-generation speed. As with ancient (DNA) evolution, so too with new techno-cultural (r)evolution: Even a fractional-percent advantage grows exponentially, resulting in a swift and complete displacement of the old.

We seek news of aging reversal and nootropics (memory and cognitive enhancers). We hunt down ways to get ahead of the FDA-EMA-CFDA curve, even risking the very youth and cognition we seek to extend. Loopholes in the global regulatory fabric include "natural" products, medical tourism, "practice of medicine" (including surgical procedures and stem-cell therapies).

Our ability to prioritize and process the news is in an auto-catalytic, positive feedback loop in which we extend our brain both biologically and electronically. Surgery could extend our brain capacity from 1.2 kg to 50 kg (routine head loads of the Sherpas of Nepal). The rate of growth of neural systems could be as fast as the doubling time of human cells (about one day) with differentiation from generic stem cells to complex neural nets recently engineered to occur in four days.

With sufficiently intimate proximity of two or more kg-scale brains, the possibility of mind-backups might be closer than via cloning (which lacks neural copying) or via computer simulation (which requires deeper understanding than mere bio-copying and has a millionfold energy inefficiency relative to brains).

The news is that we can measure and manipulate human neural development and activity with the exponentially improving "innovative neurotechnologies" (the last two letters of the BRAIN Initiative acronym). If (when) these augmentations begin to seriously help us process information, that would be mind-boggling and important news.

MEMORY IS A LABILE FABRICATION

KATE JEFFERY

Professor of behavioral neuroscience, Dept. of Psychology, University College London

We used to think of memory as a veridical record of events past, like a videotape in our heads always on hand to be replayed. Of course, we knew memory to be far more fragile and incomplete than a real videotape: We forget things, and many events aren't even stored in the first place. But when we replay our memories, we feel sure that what we do recall really happened. Indeed, our entire legal system is built on this belief.

Three scientific discoveries in the past century have changed that picture: two some time ago, and one (the "news") recent. Some time ago, we learned that memory is not a record so much as a reconstruction. We don't recall events so much as reassemble them, and crucial aspects of the original event may get substituted: It wasn't Georgina you ran into that day, it was Julia; it wasn't Monte Carlo, it was Cannes; it wasn't sunny, it was overcast (it rained later, remember)? Videotapes never do that—they get ragged and skip sections or lose information, but they don't *make things up*.

It has also been known since the 1960s that the act of reactivating a memory renders it temporarily fragile, or "labile." In its labile state, a memory is vulnerable to disruption and might be stored again in altered form. In the laboratory, this alteration is usually a degradation induced by some memory-unfriendly

411

agent, like a protein-synthesis inhibitor. We knew such drugs could affect the formation of memories, but surprisingly they can also disrupt a memory *after* it has been formed.

The story doesn't end there. Recently it has been shown that memories aren't just fragile when they've been reactivated but can actually be deliberately altered. Using some of the amazing new molecular genetic techniques developed in the past three decades, we can identify which subset of neurons participated in the encoding of an event, and then experimentally reactivate only those specific neurons, so that the animal is forced (we believe) to recall the event. During this reactivation, scientists have been able to tinker with these memories so that they end up different from the originals. So far, these tinkerings have just involved changing emotional content—such that, for example, a memory of a place which was neutral becomes positive, or a positive one becomes negative, so that the animal subsequently seeks out or avoids those places. But we aren't far from trying to write new events into these memories, and this will likely be achievable.

Why would we evolve a disconcerting system like this? Why can't memory be more like a videotape, so that we can trust it more? We don't know the answer for sure yet, but evolution doesn't care about veracity, it cares only about survival, and there's usually a good reason for apparently odd design features.

The advantages of the constructive nature of memory seem obvious: To remember every pixel of a life experience requires enormous storage capacity; it's a far more economical use of our synapses to stockpile a collection of potential memory ingredients and simply record each event in the form of a recipe: Take a pinch of a Southern French beach, add a dash of old school friend, mix in some summer weather, etc. Many theoretical neuroscientists think the labile nature of memory may allow construction of supermemories (called semantic memories)—agglomerations of individual event-memories combined to form an overarching piece of knowledge about the world. After a few visits to the Mediterranean, you learn that it's usually sunny, and so the odd incidence of overcast gloom gets washed out and fades from recollection. Your behavior thus becomes adapted not to a specific past event but to the general situation, and you know on holiday to pack sunscreen and not umbrellas.

The fabricated, labile nature of memory is at once a reason for amazement and concern. It is amazing to think that the brain is constantly reassembling our past, and that the past is not really as we think it is. It is concerning because this constructed past seems extraordinarily real—almost as real as our present—and we base our behavior on it. Thus, an eye-witness will make confident assertions that lead to someone's lifelong incarceration, and nobody worries about this except neuroscientists. It is also amazing/concerning that, as scientists and doctors, we are now on the threshold of memory editing—able to selectively alter a person's life memories.

The therapeutic potential of this is exciting—imagine being able to surgically reduce the pain of a traumatic memory! But these are technologies to use with care. In reaching into the brain and changing a person's past, we may change who they are. However, one could argue that the fabricated and labile nature of our memories means that perhaps we aren't really who we think we are anyway.

THE CONTINUALLY NEW YOU

STEPHEN M. KOSSLYN

Founding dean, Minerva Schools, Keck Graduate Institute; co-author
(with G. Wayne Miller), *Top Brain, Bottom Brain*

One of my undergraduate mentors, a senior scientist nearing
the end of a long and distinguished career, once commented
to me that even after an extraordinarily close marriage of more
than fifty years, his wife could still say and do things that surprised him. I suspect he could have extended the observation:
For better or worse, even after a lifetime of living, you can still
learn something new and surprising about yourself. Who and
what we are will always have an element of something new,
simply because of how the brain works. Here's why:

1. How we respond to objects and situations we perceive
   or ideas we encounter depends on our current cognitive
   state, wherein different concepts are "primed." Primed
   concepts are activated in our minds and influence how
   we interpret and respond to current situations. A huge
   literature now documents the effects of such priming.

2. How we interpret new stimuli or ideas relies in part on
   chaotic processes. Here's my favorite analogy for this:
   A raindrop dribbles down a windowpane. An identical
   raindrop, starting at the same spot on the window, will
   trace a different path. Even very small differences in the
   start state will affect the outcome (this is part and parcel

of what it is to be a chaotic system). The state of the windowpane, which depends on ambient temperature, effects of previous raindrops, and other factors, is like the state of the brain at a particular point in time: Depending on what one has just encountered and what one was thinking and feeling, different concepts will be primed, and this priming will influence the effects of a new perception or idea.

3. With age and experience, the structure of information stored in long-term memory becomes increasingly complex. Hence, priming has increasingly nuanced effects, which become increasingly difficult to predict.

4. In short, each of us grows as we age and experience more and varied situations and ideas, and we can never predict perfectly how we'll react to a new encounter. Why not? What we understand about ourselves depends on what we paid attention to at the time events unfolded and on our imperfect conceptual machinery for interpreting ourselves. Our understanding of ourselves will not capture the subtle effects of the patterns of priming affecting our immediate perceptions, thoughts, and feelings. Thus, although we cannot be forever young, we can be indefinitely new—at least in part.

So we should give others and ourselves some slack. We should be forgiving when friends surprise us negatively—the friends, too, may be surprised. And the same is true of us.

TODDLERS CAN MASTER COMPUTERS

ALISON GOPNIK

Psychologist, UC Berkeley; author, *The Gardener and the Carpenter: What the New Science of Child Development Tells Us About the Relationship Between Parents and Children*

In the last couple of years, toddlers and even babies have begun to be able to use computers. This may seem like the sort of minor news that shows up in the lifestyle section of the paper and in cute YouTube videos. But it actually presages a profound change in the way human beings live.

Touch and voice interfaces have become ubiquitous only recently; it's hard to remember that the iPhone is just eight years old. For grown-ups, these interfaces are a small additional convenience, but they transform the way young children interact with computers. For the first time, a toddler can directly control a smartphone or tablet.

And they do. Young children are fascinated by these devices and remarkably good at getting them to do things. In recognition of this, in 2015 the American Academy of Pediatrics issued a new report about very young children and technology. For years the Academy had recommended that children younger than two should have no access to screens at all. The new report recognizes that this recommendation has become impracticable. It focuses instead, sensibly, on ensuring that when young children look at screens, they do so in concert with attentive adults, and that adults supervise what children see.

But this isn't just news for anxious parents; it's important for the future of the human species. There is a substantial difference between the kind of learning we do as adults, or even as older children, and the kind of learning we do before we are five. For adults, learning mostly requires effort and attention; for babies, learning is automatic. Grown-up brains are more plastic than we once thought (neural connections can rewire), but very young brains are far more plastic; young children's brains are designed to learn.

In the first few years of life, we learn about the way the physical, biological, and psychological world work. Even though our everyday theories of the world depend on our experience, by the time we're adults we simply take them for granted—they're part of the unquestioned background of our lives. When technological, culturally specific knowledge is learned early, it becomes part of the background too. In our culture, children learn how to use numbers and letters before they're five. In rural Guatemala, they learn how to use a machete. These abilities require subtle and complicated knowledge, but it's a kind of knowledge that adults in the culture hardly notice (though it may startle visitors from another culture).

Until now, we couldn't assume that people would know how to use a computer in the way we assume they know how to count. Our interactions with computational systems depended on first acquiring the skills of numeracy and literacy. You couldn't learn how a computer worked without first knowing how to use a keyboard. That ensured that people learned about computers with relatively staid and inflexible old brains. We think of millennial high-school tech whizzes as precocious "digital natives." But even they only really began to learn about computers after they'd reached puberty. And that is just when brain plasticity declines precipitously.

The change in interfaces means that the next generation really will be digital natives. They will be soaked in the digital world and will learn about computers the way previous generations learned language—even earlier than previous generations learned how to read and add. Just as every literate person's brain has been reshaped by reading, my two-year-old granddaughter's brain will be reshaped by computing.

Is this a cause for alarm or celebration? The simple answer is that we don't know and we won't for at least another twenty years, when today's two-year-olds grow up. But the past history of our species should make us hopeful. After all, those powerful early learning mechanisms are exactly what allowed us to collectively accumulate the knowledge and skill we call culture. We can develop new kinds of technology as adults because we mastered the technology of the previous generation as children. From agriculture to industry, from stone tools to alphabets to printed books, we humans reshape our world, and our world reshapes our brains. Still, the emergence of a new player in this distinctively human process of cultural change is the biggest news there can be.

THE PREDICTIVE BRAIN

LISA FELDMAN BARRETT

University Distinguished Professor of psychology, Northeastern University; neuroscientist and research scientist, Psychiatry Department, Massachusetts General Hospital; lecturer in psychiatry, Harvard Medical School

Your brain is predictive, not reactive. For many years, scientists believed that your neurons spend most of their time dormant and wake up only when stimulated by some sight or sound. Now we know that all your neurons are firing constantly, stimulating one another at various rates. This intrinsic brain activity is one of the great recent discoveries in neuroscience. Even more compelling is what this brain activity represents: millions of predictions of what you will encounter next in the world, based on your lifetime of past experience.

Many predictions are at a micro level, predicting the meaning of bits of light, sound, and other information from your senses. Every time you hear speech, your brain breaks up the continuous stream of sound into phonemes, syllables, words, and ideas by prediction. Other predictions are at the macro level. You're interacting with a friend and, based on context, your brain predicts that she will smile. This prediction drives your motor neurons to move your mouth in advance to smile back, and your movement causes your friend's brain to issue new predictions and actions, back and forth, in a dance of prediction and action. If predictions are wrong, your brain has mechanisms to correct them and issue new ones.

If your brain didn't predict, sports couldn't exist. A purely reactive brain wouldn't be fast enough to parse the massive sensory input around you and direct your actions in time to catch a baseball or block a goal. You also would go through life constantly surprised.

The predictive brain will change how we understand ourselves, since most psychology experiments still assume the brain is reactive. Experiments proceed in artificial sequences called "trials," where test subjects sit passively, are presented with images, sounds, words, etc., and make one response at a time—say, by pressing a button. Trials are randomized to keep one from affecting the next. In this highly controlled environment, the results come out looking as if the subject's brain makes a rapid automatic response followed by a controlled choice about 150 milliseconds later—as if the two responses came from distinct systems in the brain. These experiments fail to account for a predicting brain, which never sits awaiting stimulation but continuously prepares multiple, competing predictions for action and perception, while actively collecting evidence to select between them. In real life, moments, or "trials," are never independent, because each brain state influences the next. Most psychology experiments are therefore optimized to disrupt the brain's natural process of prediction.

The predictive brain presents us with an unprecedented opportunity for new discoveries about how a human brain creates a human mind. New evidence suggests that thoughts, feelings, perceptions, memories, decision making, categorization, imagination, and many other mental phenomena, which historically are treated as distinct brain processes, can all be united by a single mechanism: prediction. Even our theory of human nature is up for grabs, as prediction deprives us of our most cherished narrative, the epic battle between rationality and emotions to control behavior.

A NEW IMAGING TOOL

ALUN ANDERSON

Former editor-in-chief and publishing director, *New Scientist*;
author, *After the Ice*

New tools and techniques in science don't usually garner as much publicity as big discoveries, but there's a sense in which they're much more important. Think of telescopes and microscopes: Both opened vast fields of endeavor that are still spawning thousands of major advances. And although they may not make newspaper front pages, new tools are often the biggest news for scientists—published in prestigious journals and staying at the top of citation indices for years on end. Clever tools are the long-lasting news behind the news, driving science forward for decades.

One example has just come along which I really like. A new technique makes it possible to see directly the very fast electrical activity occurring within the nerve cells of the brain of a living, behaving animal. Neuroscientists have had something like this on their wish list for years, and it's worth celebrating. The technique puts into nerve cells a special protein that can turn the tiny voltage changes of nerve activity into flashes of light. These can be seen with a microscope and recorded in exquisite detail, providing a direct window into brain activity and the dynamics of signals traveling through nerves. This is especially important because the hot news is that information contained in the nerve pulses speeding around the brain is likely coded not just in the rate at which those pulses arrive but also in their timing, with the two working at different resolutions. To start to speak neuron and thus understand our brains, we'll have to come to grips with the dynamics of signaling and relate it to what an animal is actually doing.

The new technique, developed by Yiyang Gong and colleagues in Mark Schnitzer's lab at Stanford University and published in the journal *Science*, builds on past tools for imaging nerve impulses.* One well-established method takes advantage of the calcium ions that rush into a nerve cell as a signal speeds by. Special chemicals that emit light when they interact with calcium make this electrical activity visible, but they're not fast or sensitive enough to capture the speed with which the brain works. The new technique goes further by using a rhodopsin protein (called Ace), which is sensitive to voltage changes in the nerve cell membrane, fused to another protein (mNeon), which fluoresces brightly. This imaging technique will take its place beside other recent developments extending the neuroscientist's reach. New optogenetic tools enable researchers to use light signals to switch particular nerve cells off and on to help figure out what part they play in a larger circuit.

Without constantly inventing new ways to probe the brain, the eventual goal of understanding how our 90 billion nerve cells provide us with thought and feeling will be intractable. Although we have some good insights into our cognitive strategies from psychology, deep understanding of how individual neurons work, and rapidly growing maps of brain circuitry, the vital territory in the middle—how circuits of particular linked neurons work—is tough to explore. To make progress, neuro-

---

* http://science.sciencemag.org/content/early/2015/11/18/science. aab0810.abstract

scientists dream of experiments wherein they can record what's happening in many nerves in a circuit while also switching parts of the circuit off and on and seeing the effect on a living animal's behavior. Thanks to new tools, this remarkable dream is close to coming true; when it does, the toolmakers will once again have proved that in science it's new tools that create new ideas.

SENSORS: ACCELERATING THE PACE OF SCIENTIFIC DISCOVERY

PAUL SAFFO

Technology forecaster; consulting associate professor,
Stanford University

Behind every great scientific discovery is an instrument. From Galileo and his telescope to Arthur Compton and the cloud chamber, our most important discoveries are underpinned by device innovations that extend human senses and augment human cognition. This is a crucially important science-news constant, because without new tools discovery would slow to a crawl. Want to predict the next big science surprise a decade from now? Look for the fastest-moving technologies and ask what new tools they enable.

For the last half-century, digital technology has delivered the most powerful tools, in the form of processors, networking, and sensors. Processing came first, providing the brains for space probes and the computational bulldozers needed for tackling computation-intensive research. Then with the advent of the Arpanet, the Internet, and the World Wide Web, networking became a powerful medium for accessing and sharing scientific knowledge—and connecting remotely to everything from supercomputers to telescopes. But it is the third category—sensors, and an even newer category of robust effectors—that is poised to accelerate and utterly change research and discovery in the decades ahead.

First, we created our computers, then we networked them,

and now we are giving them sensory organs to observe—and manipulate—the physical world in the service of science. And thanks to the phenomenon described by Moore's Law, sensor cost/performance is racing ahead as rapidly as chip performance. Ask any amateur astronomer: For a few thousand dollars, they can buy digital cameras that were beyond the reach of observatories a decade ago.

The entire genomics field owes its very existence and future to sensors. Craig Venter's team became the first to decode the human genome in 2001, by leveraging computational power and sensor advances to create a radically new and radically less expensive sequencing process. Moreover, the cost of sequencing is already dropping more rapidly than the curve of Moore's Law. Follow out the Carlson Curve (as the sequencing price/performance curve was dubbed by *The Economist*), and the cost of sequencing a genome is likely to plummet below $1.00 well before 2030. Meanwhile, the gene editing made possible by the CRISPR/Cas9 system is possible only because of ever more powerful and affordable sensors and effectors. Just imagine the science possible when sequencing a genome costs a dime and networked sequencing labs-on-a-chip are cheap enough to be tossed out and discarded like RFID tags.

Sensors and digital technology are also driving physics discovery. The heart of CERN's Large Hadron Collider is the CMS detector, a 14,000-ton assemblage of sensors and effectors that has been dubbed "science's cathedral." Like a cathedral of old, it is served by nearly 4,000 people drawn from over forty countries, and it's so popular that a scientific journal featured a color-in centerfold of the device in its year-end issue.

Sensors are also opening vast new windows on the cosmos. Thanks to the relentless advance of sensors and effectors in the form of adaptive optics, discovery of extrasolar planets moved from science fiction to commonplace with breathtaking speed. In the near future, sensor advances will allow us to analyze exoplanetary atmospheres and look for signatures of civilization. The same trends will open new horizons for amateur astronomers, who will soon enjoy affordable technical means to match the *Kepler* space telescope in planet-finding prowess. Sensors are thus as much about democratizing amateur science as the creation of ever more powerful instruments. The *Kepler* satellite imaged a field of 115°, or a mere 0.25 percent of the sky. Planet-finding amateurs wielding digitally empowered backyard scopes could put a serious dent in the 99.75 percent of the sky yet to be examined.

Another recent encounter between amateurs and sensors offers a powerful hint of what is to come. Once upon a time, comets were named after human discoverers, and amateurs hunted comets with such passion that more than one would-be comet hunter relocated eastward in order to get an observing jump on the competition. Now, comets have names like 285P/ Linear, because robotic systems are doing the discovering, and amateur comet-hunting is in steep decline. Amateurs will find other things to do (like search for planets), but it's hard not to feel a twinge of nostalgia for a lost time when that wispy apparition across the sky carried a romantic name like Hale-Bopp or Ikeya-Seki rather than C/2011-L4 PanStarrs.

This shift in cometary nomenclature hints at an even greater sea change to come in the relationship between instrument and discoverer. Until now, the news has been of ever more powerful instruments created in the service of amplifying human-driven discovery. But just as machines today are better comet finders than humans, we are poised on the threshold of a time when machines do not merely amplify but displace the human researcher. When that happens, the biggest news of all will be when a machine wins a Nobel Prize alongside its human collaborators.

3D PRINTING IN THE MEDICAL FIELD

SYED TASNIM RAZA

Medical director, Cardiac Surgery Step-Down Unit, Columbia
University Medical Center and New York–Presbyterian University
Hospital

Within the field of medicine, arguably the most progress made in
the last few decades is in clinical imaging, starting with simple X-
rays and moving to such current technologies as CAT scans and
fMRI. And then there is ultrasonography, extensively used in
diagnostic and therapeutic interventions (such as amniocentesis
during pregnancy) and imaging of the heart (echocardiography).
Cardiologists have used various other imaging modalities for diagnosis of heart conditions, including heart catheterization. They
perform contrast studies by injecting radio-opaque material into
the heart chambers or blood vessels while recording moving
images (angiograms). And then there is Computed Tomographic
Angiography of the heart (CTA), with its 3D reconstruction,
providing detailed information of the cardiac structure.

Now comes 3D printing, adding another dimension to the
imaging of the human body. In its current form, using CAD
(computer-aided design programs), engineers develop a three-
dimensional computer model of any object to be "printed" (or
built), which is then translated into a series of two-dimensional
slices of the object. The 3D printer can then lay down thousands
of layers until the vertical dimension is achieved and the object
is built.

Within the last few years, this technology has been used in the medical field, particularly in surgery. In cardiac surgery, 3D printing is applied mostly in congenital heart disease. In congenital heart malformations, many variations from the normal can occur. With current imaging techniques, surgeons have a fair idea what to expect before operating, but many times they have to "explore" the heart during surgery to ascertain the exact malformation and then plan at the spur of the moment. With the advent of 3D printing, they can do a CTA scan of the heart, with its three-dimensional reconstruction, which can then be fed into the 3D printer, creating a model of the malformed heart. The surgeons can then study this model and even cut slices into it to plan the exact operation they will perform, saving valuable time during the procedure itself.

Three-dimensional printing is used in many areas of medicine, particularly in orthopedics. One of the more exciting areas is in building live organs for replacements, using living cells and stem cells layered onto scaffolding of the organ to be "grown," so that the cells can grow into skin, earlobe, or other organs. Someday organs may be grown for each individual from his or her own stem cells, obviating the risk of rejection and avoiding poisonous anti-rejection medicines. Exciting development.

DEEP SCIENCE

BRIAN KNUTSON
Associate professor of psychology and neuroscience,
Stanford University

The decade of the brain is maturing into the century of the mind. New bioengineering techniques can resolve and perturb brain activity with unprecedented specificity and scope (including neural control with optogenetics, circuit visualization with fiber photometry, receptor manipulation with DREADDs, gene sculpting with CRISPR/Cas9, and whole brain mapping with CLARITY). These technical advances have captured well-deserved media coverage and inspired support for brain-mapping initiatives. But conceptual advances are also needed. We might promote faster progress by complementing existing "broad science" initiatives with "deep science" approaches able to bridge the chasms separating different levels of analysis. Thus, some of the most interesting neuroscientific news on the horizon might highlight not only new scientific content (for example, tools and findings) but also new scientific approaches (for example, deep- versus broad-science approaches).

What is "deep science"? Deep-science approaches seek first to identify critical nodes (or units) within different levels of analysis and determine whether they share a link across those levels. If such a connection exists, then perturbing the lower-level node could causally influence the higher-level node. Some examples of deep-science approaches include using optogenetic stimulation to alter behavior or using fMRI activity to predict psychiatric symptoms. Because deep science first seeks to link different levels of analysis, it often requires collaboration of at least two experts at those levels.

The goals of deep science stand in contrast to those of broad science. Broad science seeks to map all nodes within a level of analysis as well as links between them (for example, all neurons and their connections in a model organism, like a worm). Comprehensive characterization is a necessary step in mapping the landscapes of new data produced by novel techniques. Examples of broad-science approaches include connectomic attempts to characterize all brain cells in a circuit, or computational efforts to digitally model all circuit components. Broad-science initiatives implicitly assume that by fully characterizing a single level of analysis, a better understanding of higher-order functions will emerge. Thus, a single expert at one level of analysis can advance through persistent application of relevant methods.

Due to more variables, methods, and collaborators, deep-science approaches pose greater coordination challenges than broad-science approaches. Which nodes to target or levels to link might not be obvious at first and might require many rounds of research. Although neuroscientists have long distinguished different levels of analysis, they have often emphasized one level to the exclusion of others, or assumed that links across levels were arbitrary and thus unworthy of study. New techniques, however, have raised possibilities for testing links across levels. Thus, one deep-science strategy might involve targeting links that causally connect ascending levels of analysis. For instance, recent evidence indicates that optogenetic stimulation of midbrain dopamine neurons (the hardware level) increases fMRI activity in the striatum (the process level), which predicts approach behavior (the goal level) in rats and humans.

While deep-science findings are not yet news, I predict they soon will be. Deep science and broad science are necessary complements, but broad-science approaches currently dominate. By linking levels of analysis, deep-science approaches may more rapidly translate basic neuroscience knowledge into behavioral applications and healing interventions—which should be good news for all.

A WORLD THAT COUNTS

ALEX (SANDY) PENTLAND
Toshiba Professor of Media, Arts, and Sciences, MIT; director, MIT
Human Dynamics Lab and the Connection Science program; author,
Social Physics

In 2014 a group of Big Data scientists (including myself), repre-
sentatives of Big Data companies, and the heads of National Statistical Offices from nations in both the Northern and Southern
Hemispheres met at United Nations headquarters and plotted
a revolution. We proposed that all of the nations of the world
begin to measure poverty, inequality, injustice, and sustainability in a scientific, transparent, accountable, and comparable
manner. Surprisingly, this proposal was approved by the U.N.
General Assembly in 2015, as part of the 2030 Sustainable Development Goals.

This apparently innocuous agreement is informally known
as the Data Revolution within the U.N., because for the first
time there is an international commitment to discover and tell
the truth about the state of the human family as a whole. Since
the beginning of time, most people have been isolated, invisible to government, and without information about or input to
governmental health, justice, education, or development policies.
But in the last decade this has changed. As our U.N. Data Revolution report, titled *A World That Counts*, states:

> Data are the lifeblood of decision-making and the raw ma-
> terial for accountability. Without high-quality data providing

the right information on the right things at the right time, designing, monitoring and evaluating effective policies becomes almost impossible. New technologies are leading to an exponential increase in the volume and types of data available, creating unprecedented possibilities for informing and transforming society and protecting the environment. Governments, companies, researchers and citizen groups are in a ferment of experimentation, innovation and adaptation to the new world of data, a world in which data are bigger, faster and more detailed than ever before. This is the data revolution.

More concretely, the vast majority of humanity now has a two-way digital connection that can send voice, text, and, most recently, images and digital sensor data, because cell-phone networks have spread nearly everywhere. Information is suddenly potentially available to everyone. The Data Revolution combines this enormous new stream of data about human life and behavior with traditional data sources, enabling a new science of "social physics," which can let us detect and monitor changes in the human condition and provide precise, nontraditional interventions to aid human development.

Why would anyone believe that anything will actually come from a U.N. General Assembly promise that the National Statistical Offices of the member nations will measure human development openly, uniformly, and scientifically? It is not because anyone hopes that the U.N. will manage or fund the measurement process. Instead, we believe that uniform scientific measurement of human development will happen because international development donors are finally demanding scientifically sound data to guide aid dollars and trade relationships.

Moreover, once reliable data about development start becoming familiar to business people, supply chains and private investment will begin paying attention. A nation with poor measures of justice or inequality normally also has higher levels of corruption, and a nation with a poor record in poverty or sustainability normally also has a poor record of economic stability. As a consequence, nations with low measures of development are less attractive to business than nations with similar costs but better human-development numbers.

Historically we have been blind to the living conditions of the rest of humanity; violence or disease could spread to pandemic proportions before the news would make it to the ears of central authorities. We are now beginning to be able to see the condition of all of humanity with unprecedented clarity. Never again should it be possible to say "We didn't know." No one should be invisible. This is the world we want—a world that counts.

PROGRAMMING REALITY

NEIL GERSHENFELD
Physicist; director, MIT's Center for Bits and Atoms;
author, *The Nature of Mathematical Modeling*

The most notable scientific news story in 2015 was not obviously about science. What was apparent was the coverage of diverging economic realities. Much of the world struggled with income inequality, persistent unemployment, stagnant growth, and budgetary austerity, amid corporate profit records and a growing concentration of wealth. In turn, this gulf led to a noisy emergence of far-right and far-left political movements, offering a return to a promised better time decades (or centuries) ago. And these drove the appearance of a range of conflicts, connected by a common thread of occurring in failing and failed economies.

So what do all these dire news stories have to do with science? They share an implicit syllogism so obvious it's never mentioned: Opportunity comes from creating jobs because jobs create income, and inequality is due to the lack of income. That's what's no longer true. The unseen scientific story is to break the historical relationship between work and wealth by removing the boundary between the digital and physical worlds.

Some discoveries arrive as an event, like the flash of a lightbulb; some are best understood in retrospect as the accumulation of a body of work, where the advance is to take it seriously. This is one of those. Digitizing communication and computation required a few decades each, leading to a revolution in how knowledge is created and shared. The coverage now of 3D printing and the maker movement is only the visible tip of a much bigger iceberg, digitizing not just design descriptions for computer-controlled manufacturing machines (which is decades old) but also the designs themselves by specifying the assembly of digital materials.

Life is based on a genetic code that determines the placement of twenty standard amino acids; that was discovered (by molecular biology) a few billion years ago. We're now learning how to apply this insight beyond molecular biology; emerging research is replacing processes that continuously deposit or remove materials with ones that code the reversible construction of discrete building blocks. This is being done across disciplines and length scales, from atomically precise manufacturing to whole-genome synthesis of living cells to the three-dimensional integration of functional electronics to the robotic assembly of modular aircraft and spacecraft. Taken together, these add up to programming reality—turning data into things and things into data.

Returning to the news stories from 2015: Going to work commonly means leaving home to travel to somewhere you don't want to be, to do something you don't want to do, producing something for someone you'll never see, to get money to pay for something you want. What if you could instead just *make* what you want? In the same way that digitizing computing turned information into a commodity, digitizing fabrication reduces the cost of producing something to the incremental cost of its raw materials.

In the largest-ever gathering of heads of state, the Sustainable Development Goals were launched at the U.N. in 2015. These target worthy aims, including ending poverty and hunger, ensuring access to healthcare and energy, building infrastructure, and reducing inequality. Left unsaid is how to accomplish these goals, which will require spending vast amounts of money. But development needn't recapitulate the Industrial Revolution; just as developing countries have skipped over landlines and gone right to mobile phones, mass manufacturing with global supply chains can be replaced with sustainable local on-demand fabrication of all the ingredients of a technological civilization. This is a profound challenge, but it's one with a clear research roadmap and is the scientific story behind the news.

POINTING IS A PREREQUISITE FOR LANGUAGE

N. J. ENFIELD

Professor of linguistics, University of Sydney; research associate,
Language and Cognition Group, Max Planck Institute for
Psycholinguistics, Nijmegen, The Netherlands;
author, *Natural Causes of Language*

Research in developmental and comparative psychology has discovered that the humble pointing gesture is a key ingredient of the capacity for developing and using human language, and indeed for the very possibility of human social interaction.

Pointing gestures seem simple. We use them all the time. I might point when I give someone directions to the station, when I indicate which loaf of bread I want to buy, or when I show you where you have spinach stuck in your teeth. We often accompany such pointing gestures with words, but for infants who cannot yet talk, these gestures can work on their own.

Infants begin to communicate by pointing at about nine months of age; it's a year before they can produce even the simplest sentences. Careful experimentation has established that prelinguistic infants can use pointing gestures to ask for things, to help others by pointing things out to them, and to share experiences with others by drawing attention to things they find interesting and exciting.

Pointing does not just manipulate the other's focus of attention; it momentarily unites two people through a shared focus on something. With pointing, we do not just look at the same thing, we look at it *together*. This is a particularly human trick, and arguably what ultimately makes social and cultural institutions possible. Being able to point and to comprehend the pointing gestures of others is crucial to achieving "shared intentionality," the ability to build relationships through the sharing of perceptions, beliefs, desires, and goals.

Comparative psychology finds that pointing (in its full-blown form) is unique to our species. Few nonhuman species seem able to comprehend pointing (notably, domestic dogs can follow pointing, while our closest relatives among the great apes cannot), and there is little evidence of pointing occuring spontaneously between members of any species other than our own. Apparently only humans have the social-cognitive infrastructure needed to support the kind of cooperative and prosocial motivations that pointing gestures presuppose.

This suggests a new place to look for the foundations of human language. While research on language in cognitive science has long focused on its logical structure, the news about pointing suggests an alternative: that the essence of language is found in our capacity for the communion of minds through shared intentionality. At the center of it is the deceptively simple act of pointing, an act that must be mastered before language can be learned at all.

MACRO-CRIMINAL NETWORKS

EDUARDO SALCEDO-ALBARÁN
Philosopher; director, Scientific Vortex, Inc.

Powerful computation today boosts our ability to perceive and understand the world. The more data we process and analyze, the more natural and social phenomena we discover and understand. Copious social data reveal global trends. For instance, analyzing masses of judicial information with current computational tools has exposed a new and complex social phenomenon: macro-criminal networks.

Our brains make sense of those social networks in which only about 150 to 200 individuals participate. Known as "Dunbar's number," this is an approximation of the social-network size we can interact with. Thus macro-criminal networks cannot be perceived or analyzed without computational power, algorithms, and the right concepts of social complexity.

Unfortunately, we as a society lack tools, legislation, and enforcement mechanisms to confront these global, resilient, and decentralized structures, which are characterized by messy hierarchies and various types of leaders. Macro-criminal networks overwhelm most law-enforcement agents, who still search for criminal organizations with simple hierarchies run by full-time criminals and commanded by a single boss. This classic idea of "organized crime" is outdated and doesn't reflect the complexity of the macro-criminal networks now being uncovered.

Investigating and prosecuting crime today without the right concepts or computational tools for processing and analyzing enormous amounts of data is like studying galaxies with 17th-century telescopes. The greatest challenge in confronting macro-criminal networks is not the adoption of powerful computers or the application of deep learning but modifying that mindset of scholars, investigators, prosecutors, and judges. Legislation focused on one victim and one victimizer leads to wrong analysis and insufficient enforcement of such systemic crimes as the corruption in Latin America and West Africa, human trafficking in Eastern Europe, and forced displacement in Central Africa. As a consequence, structures supporting those crimes worldwide are overlooked.

Crime in its various expression is always news. From corruption to terrorism and trafficking activities, crime affects our way of life while hampering development in various countries. Understanding the phenomenon of huge, resilient, and decentralized criminal macro-structures is critical for achieving global security. We need to commit and allocate the right scientific, institutional, and economic resources to deal with it.

VIRTUAL REALITY GOES MAINSTREAM

THOMAS METZINGER

Philosopher, Johannes Gutenberg-Universität Mainz;
editor, Open-Mind.net; author, *The Ego Tunnel*

Suppose you have just popped one of those new hedonic enhancement pills for virtual environments. Not the dramatic, illegal stuff, just the legal pharmaceutical enhancement that comes as a direct-to-consumer advertising gift with the gadget itself. It has the great advantage of blocking nausea and thereby stabilizing the realtime, fMRI-based neurofeedback-loop into your own virtual reality (allowing you to interact with the unconscious causes of your own feelings directly, as if they were now part of an external environment), while at the same time nicely minimizing the risk of depersonalization disorder and *Truman Show* delusion. These pills also reliably prevent addiction and the diminished sense of agency upon reentering the physical body following longtime immersion—at least the package leaflet says so. As you turn on the device, two of your "Selfbook-friends" are already there, briefly flashing their digital subject identifiers. Their avatars immediately make eye contact and smile at you, and you automatically smile back, while you feel the pill taking effect. Fortunately, they can see neither the new Immersive Porn trial version nor the expensive avatar that represents your Compassionate Self. You only use that twice a week in your psychotherapy sessions. The NSA, however, sees everything.

In 2016, VR will finally break through at the mass-consumer level. Moreover, users will soon be able to toggle between virtual, augmented, and substitutional reality, experiencing virtual elements intermixed with their "actual" physical environment, or an omnidirectional video feed giving them the illusion of being in a different location in space and/or time. Oculus Rift, Zeiss VR One, Sony PlayStation VR, HTC Vive, Samsung's Galaxy Gear VR or Microsoft's HoloLens are just the beginning, and it's hard to predict the psychosocial consequences over the next two decades as an accelerating technological development is driven by massive market forces and not by scientists anymore. There will be great benefits (just think of the clinical applications) and a host of new ethical issues, ranging from military applications to data protection (for example, "kinematic fingerprints" generated by motion-capture systems, avatar ownership, and individuation will become important questions for regulatory agencies to consider).

The *real* news, however, may be that the general public will gradually acquire a new and intuitive understanding of what their conscious experience really is and what it always has been. VR is the representation of *possible* worlds and *possible* selves, with the aim of making them appear real—ideally, by creating a subjective sense of "presence" in the user. Interestingly, some of our best theories of the human mind and conscious experience describe such experience in a similar way. Leading theoretical neurobiologists, like Karl Friston, and eminent philosophers, like Jakob Hohwy and Andy Clark, describe it as the constant creation of internal models of the world, virtual neural representations of reality which express probability density functions and work by continuously generating hypotheses about the hidden causes of sensory input, minimizing their prediction error. In 1995, Finnish philosopher Antti Revonsuo pointed out that conscious experience *is* a virtual model of the world—a dynamic internal simulation which in standard situations cannot be experienced as a virtual model because it is phenomenally transparent: We "look through it" as if we were in direct and immediate contact with reality. What is historically new, and what creates not only novel psychological risks but also entirely new ethical and legal dimensions, is that one virtual reality gets ever more deeply embedded into another virtual reality. The conscious mind, which has evolved under specific conditions and over millions of years, now gets causally coupled and informationally woven into technical systems for representing possible realities. Increasingly, consciousness is not only culturally and socially embedded but also shaped by a specific technological niche that, over time, acquires rapid, autonomous dynamics and new properties. This creates a complex convolution, a nested form of information flow in which the biological mind and its technological niche influence each other in ways we are just beginning to understand.

THE TWIN TIDES OF CHANGE

TIMO HANNAY
Founder, SchoolDash; managing director, Digital Science
(Macmillan); co-organizer, Sci Foo Camp

News stories are by their nature ephemeral. Whipped up by the media (whether mass or social), they soon dissipate, like ripples on the surface of the sea. More significant and durable are the great tides of social change and technological progress on which they ride. It is these that will continue to matter for generations to come. Fortunately, like real tides, they tend to be predictable.

One such inexorable trend is our changing relationship with the natural world—most vividly represented by the ongoing debate about whether humanity's impact has been so profound as to justify the christening of a new geological epoch: the Anthropocene. Whether or not a consensus emerges in the next few years, it will do so eventually, for our effect on the planet will only grow. This is in part because our technological capabilities continue to expand, but an even more important driver is our evolving collective psyche.

Since Darwin showed us that we are products of the natural world rather than its divinely appointed overlords, we've been reluctant to fully impose our will, fearful of our own omnipotence and concerned that we'll end up doing more harm than good. But we might as well be trying, Canute-like, to hold back the sea. For whatever is beyond the pale today will eventually come to seem so natural that it will barely register as news—even if it takes the death of the old guard to usher in a new way of thinking. To future generations, genetic engineering of plants and animals (and humans) will seem as natural as selective breeding is today, and planetary-scale geoengineering will become as necessary and pervasive as the construction of dams and bridges.

As for our place in nature, so too for our relationship with technology. Recent progress in artificial intelligence and bionics, in particular, have led to a lot of soul-searching about who—or what—is in charge, and even what it means to be human. The Industrial Revolution saw machines replace human physical labor, but now that they are replacing mental labor too, what's left for people to do? Even those who don't fear for their jobs might be angry when they discover that their new boss is an algorithm.

Yet since the invention of the wheel, humans have lived in happy and productive symbiosis with their technologies. Despite our ongoing appetite for scare stories, we'll continue to embrace innovations as the primary source of improvements in our collective well-being. In doing so, we'll come to see them as natural extensions of ourselves—indeed, as enablers and enhancers of our humanity—rather than as something artificial or alien. A life lived partly in virtual reality will be no less real than one seen through a pair of contact lenses. Someone with a computer inserted in their brain rather than merely in their pocket will be seen as no less human than someone with a pacemaker. And traveling in a vehicle or aircraft without a human at the controls will be seen not as reckless but as reassuring. We are surely not too far from the day when *Edge* will receive its first contribution from a genetically enhanced author or an artificial intelligence. That, too, will be big news, but not for long.

Thus, humanity is subject to two inexorably rising tides: a scientific and technological one in which the magical eventually becomes mundane, and a psychological and social one in which the unthinkable becomes unremarkable. News stories will come and go like breaking waves; meanwhile, beneath them, inconspicuous in their vastness, the twin tides of technological and social change will continue their slow but relentless rise, testing and extending the boundaries of human knowledge and acceptance. This will be the real story of our species and our age.

IMAGING DEEP LEARNING

ANDY CLARK

Philosopher and cognitive scientist, University of Edinburgh;
author, *Surfing Uncertainty*

The world is increasingly full of deep architectures—multilevel artificial neural networks employed to discover (via deep learning) patterns in large data sets, such as images and texts. But the power and prevalence of these deep architectures mask a major problem—the problem of knowledge opacity. Such architectures learn to do wonderful things, but they do not (without further coaxing) reveal just what knowledge they are relying upon when they do them.

This is both disappointing (theoretically) and dangerous (practically). Deep learning and the patterns it extracts now permeate every aspect of our daily lives, from online search and recommendation systems to bank-loan applications, healthcare, and dating. Systems that have that much influence over our destinies ought to be as transparent as possible. The good news is that new techniques are emerging to probe the knowledge gathered and deployed by deep-learning systems.

In June 2015, Alexander Mordvintsev and co-authors published online a short piece entitled "Inceptionism: Going Deeper into Neural Networks." Named after a specific architecture, "Inceptionism" was soon trending on just about every geeky blog in the universe. The authors took a trained-up network capable of deciding what is shown in a given image. They then devised an automatic way to get the network to enhance an input image in ways that would tweak it toward an image that would be classified, by that network, as some specific item. This involved essentially running the network in reverse (hence the frequent references to "networks dreaming" and "reverse hallucination" in the blogs). For example, starting with random noise and a target classification, while constraining the network to respect the statistical profiles of the real images it had been trained on, the result would be a vague, almost impressionistic image that reveals how the network thinks that kind of item ("banana," "starfish," "parachute," or whatever) should look.

There were surprises. The target "barbell," for example, led the network to hallucinate two-ended weights all right—but every barbell still had the ghostly outline of a muscular arm attached. That tells us that the network has not quite isolated the core idea yet, though it got pretty close. Most interesting, you can now feed in a real image, pick one layer of your multilevel network, and ask the system to enhance whatever is detected. This means you can use inceptionism to probe and visualize what's going on at each processing layer. Inceptionism is thus a tool for looking into the network's multilevel mind, layer by layer.

Many of the results were psychedelic—repeated enhancements at certain levels resulted in images of fractal beauty, mimicking trippy artistic forms and motifs. This was because repeating the process results in feedback loops. The system is (in effect) being asked to enhance whatever it sees in the image as processed at some level. So if it sees a hint of birdiness in a cloud, or a hint of faceness in a whirlpool, it will enhance that, bringing out a little more of that feature or property. If the resulting enhanced image is then fed in as input, and the same technique applied, those enhancements make the hint of birdiness (or whatever) even stronger, and another round of enhancement ensues. This rapidly results in some image elements morphing toward repeating, dreamlike versions of familiar things and objects.

If you haven't yet seen these fascinating images, you can check them out online in the "inceptionism gallery," and even create them using the code available in DeepDream. Inceptionist images turn out to be objects of beauty and contemplation in their own right, and the technique may thus provide a new tool for creative exploration—not to mention suggestive hints about the nature of our own creative processes. But this is not just, or even primarily, image-play. Such techniques are helping us understand what kinds of things these opaque, multilevel systems know—what they rely upon layer by layer as their processing unfolds.

This is neuroimaging for the artificial brain.

THE NEURAL NET RELOADED

JAMSHED BHARUCHA

Psychologist; president emeritus, Cooper Union

The neural network has been resurrected. After a troubled sixty-year history, it has crept into the daily lives of hundreds of millions of people, in the span of just three years.

In May 2015, Sundar Pichai announced that Google had reduced errors in speech recognition to 8 percent, from 23 percent only two years earlier. The key? Neural networks, rebranded as deep learning. Google reported dramatic improvements in image recognition just six months after acquiring DNN Research, a startup founded by Geoffrey Hinton and two of his students. Back-propagation is back—with a Big Data bang. And it's suddenly worth a fortune.

The news wasn't on the front pages. There was no scientific breakthrough. Nor was there a novel application. Why is it news? The scale of the impact is astonishing, as is the pace at which it was achieved. Making sense of noisy, infinitely variable, visual and auditory patterns has been a Holy Grail of artificial intelligence. Raw computing power has caught up with decades-old algorithms. In just a few short years, the technology has leapt from laboratory simulations of oversimplified problems to cell-phone apps for the recognition of speech and images in the real world.

Theoretical developments in neural networks have been mostly incremental since the pioneering work on self-organization in the 1970s and back-propagation in the 1980s.

The tipping point was reached recently not by fundamentally new insights but by processing speeds that make possible larger networks, bigger data sets, and more iterations.

This is the second resurrection of neural networks. The first was the discovery by Hinton and Yann LeCun that multilayered networks can learn nonlinear classification. Before this breakthrough, Marvin Minsky and Seymour Papert had all but decimated the field with their 1969 book *Perceptrons*. Among other things, they proved that Frank Rosenblatt's perceptron could not learn classifications that are nonlinear.

Rosenblatt developed the perceptron in the 1950s. He built on foundational work in the 1940s by McCulloch and Pitts, who showed how patterns could be handled by networks of neurons, and Donald Hebb, who hypothesized that the connection between neurons is strengthened when connected neurons are active. The buzz created by the perceptron can be relived by reading "ELECTRONIC 'BRAIN' TEACHES ITSELF," in the July 13, 1958, *New York Times*. The *Times* quoted Rosenblatt as saying that the perceptron "will grow wiser as it gains experience," adding that "the Navy said it would use the principle to build the first Perceptron 'thinking machines' that will be able to read or write."

Minsky and Papert's critique was a major setback, if not a fatal one, for Rosenblatt and neural networks. But a few people persisted quietly, among them Stephen Grossberg, who began working on these problems while an undergraduate at Dartmouth in the 1950s. By the 1970s, Grossberg had developed an unsupervised (self-organizing) learning algorithm that balanced the stability of acquired categories with the plasticity necessary to learn new ones.

Hinton and LeCun addressed Minsky and Papert's challenge

and brought neural nets back from obscurity. The excitement about back-propagation drew attention to Grossberg's model, as well as to the models of Fukushima and Kohonen. But in 1988, Steven Pinker and Alan Prince did to neural nets what Minsky and Papert had done two decades earlier, with a withering attack on the worthiness of neural nets for explaining the acquisition of language. Once more, neural networks faded into the background.

After Hinton and his students won the ImageNet challenge in 2012, with a quantum improvement in performance on image recognition, Google seized the moment, and neural networks came alive again.

The opposition to deep learning is gearing up already. All methods benefit from powerful computing, and traditional symbolic approaches also have demonstrated gains. Time will tell which approaches prevail, and for what problems. Regardless, 2012–2015 will have been the time when neural networks placed artificial intelligence at our fingertips.

DIFFERENTIABLE PROGRAMMING

DAVID DALRYMPLE

Computer scientist, neuroscientist; research affiliate, MIT Media Lab

Over the past few years, a raft of classic challenges in artificial intelligence which stood unmet for decades were overcome almost without warning, through an approach long disparaged by AI purists for its "statistical" flavor: It was essentially about learning probability distributions from large volumes of data rather than examining humans' problem-solving techniques and attempting to encode them in executable form. The formidable tasks it has solved range from object classification and speech recognition to generating descriptive captions for photos and synthesizing images in the style of famous artists—even guiding robots to perform tasks for which they were never programmed!

This newly dominant approach, originally known as "neural networks," is now branded "deep learning," to emphasize a qualitative advance over the neural nets of the past. Its recent success is often attributed to the availability of larger data sets and more powerful computing systems, or to large tech companies' sudden interest in the field. These increasing resources have indeed been critical ingredients in the rapid advancement of the state of the art, but big companies have always thrown resources at a wide variety of machine-learning methods. Deep learning in particular has seen unbelievable advances; many other methods have also improved, but to a far lesser extent.

So what is the magic that separates deep learning from the rest and can crack problems for which no group of humans has ever been able to program a solution? The first ingredient, from the early days of neural nets, is a timeless algorithm, rediscovered again and again, known in this field as "back-propagation." It's really just the chain rule—a simple calculus trick—applied in an elegant way. It's a deep integration of continuous and discrete math, enabling complex families of potential solutions to be autonomously improved with vector calculus.

The key is to organize the template of potential solutions as a directed graph (e.g., from a photo to a generated caption, with many nodes in between). Traversing this graph in reverse enables the algorithm to automatically compute a "gradient vector," which directs the search for increasingly better solutions. You have to squint at most modern deep-learning techniques to see any structural similarity to traditional neural networks; behind the scenes, this back-propagation algorithm is crucial to both old and new architectures.

But the original neural networks using back-propagation fall far short of newer deep-learning techniques, even using today's hardware and data sets. The other key piece of magic in every modern architecture is another deceptively simple idea: Components of a network can be used in more than one place at the same time. As the network is optimized, every copy of each component is forced to stay identical (this idea is called "weight-tying"). This enforces an additional requirement on weight-tied components: They must learn to be useful in many places all at once, not specialize to a particular location. Weight-tying causes the network to learn a more generally useful function, since a word might appear at any location in a block of text, or a physical object might appear at any place in an image.

Putting a generally useful component in many places of a network is analogous to writing a function in a program and calling it in multiple spots—an essential concept in a different area of computer science, functional programming. This is more than just an analogy: Weight-tied components are actually the same concept of reusable function as in programming. And it goes even deeper! Many of the most successful architectures of the past couple of years reuse components in exactly the same patterns of composition generated by common "higher-order functions" in functional programming. This suggests that other well-known operators from functional programming might be a good source of ideas for deep-learning architectures.

The most natural playground for exploring functional structures trained as deep-learning networks would be a new language that can run back-propagation directly on functional programs. As it turns out, hidden in the details of implementation, functional programs are compiled into a computational graph similar to what back-propagation requires. The individual components of the graph need to be differentiable too, but Grefenstette *et al.* recently published differentiable constructions of a few simple data structures (stack, queue, and deque), which suggests that further differentiable implementations are probably just a matter of clever math. More work in this area may open up a new programming paradigm—differentiable programming. Writing a program in such a language would be like sketching a functional structure with the details left to the optimizer; the language would use back-propagation to automatically learn the details according to an objective for the whole program—just like optimizing weights in deep learning but with functional programming as a more expressive generalization of weight-tying.

Deep learning may look like another passing fad, in the vein of "expert systems" or "Big Data." But it's based on two timeless ideas—back-propagation and weight-tying—and although differentiable programming is a new concept, it's a natural extension of those ideas which may prove timeless itself. Even as specific implementations, architectures, and technical phrases go in and out of fashion, these core concepts will continue to be essential to the success of AI.

DEEP LEARNING, SEMANTICS, AND SOCIETY

STEVE OMOHUNDRO

Scientist, Possibility Research, Self-Aware Systems; cofounder, Center for Complex Systems Research

Deep-learning neural networks are the most exciting recent technological and scientific development. Technologically, they are soundly beating competing approaches in a wide variety of contests including speech recognition, image recognition, image captioning, sentiment analysis, translation, drug discovery, and video-game performance. This has led to huge investments by the big technology companies and the formation of more than 300 deep-learning startups with more than $1.5 billion of investment.

Scientifically, these networks are shedding new light on one of the most important scientific questions of our time: "How do we represent and manipulate meaning?" Many theories of meaning have been proposed that involve mapping phrases, sounds, or images into logical calculi with formal rules of manipulation. For example, Montague semantics tries to map natural-language phrases into a typed Lambda calculus.

The deep-learning networks naturally map input words, sounds, or images into vectors of neural activity. These vector representations exhibit a curious "algebra of meaning." For example, after training on a large English language corpus, Tomas Mikolov's Word2Vec exhibits this strange relationship: "King -

Man + Woman = Queen." His network tries to predict words from their context (or vice versa). The shift of context from "The king ate his lunch" to "The queen ate her lunch" is the same as from "The man ate his lunch" to "The woman ate her lunch." The statistics of many similar sentences lead to the vector from "king" to "queen" being the same as from "man" to "woman." It also maps "prince" to "princess," "hero" to "heroine," and many other similar pairs. Other "meaning equations" include "Paris - France + Italy = Rome," "Obama - USA + Russia = Putin," "Architect - Building + Software = Programmer." In this way, these systems discover important relational information purely from the statistics of training examples.

The success of these networks can be thought of as a triumph of distributional semantics, first proposed in the 1950s. Meaning, relations, and valid inference all arise from the statistics of experiential contexts. Similar phenomena were found in the visual domain in Radford, Metz, and Chintala's deep networks for generating images. The vector representing a smiling woman minus the woman with a neutral expression plus a neutral man produces an image of the man smiling. A man with glasses minus the man without glasses plus a woman without glasses produces an image of the woman with glasses.

Deep-learning neural networks now have hundreds of important applications. A classical challenge for industrial robots is to use vision to find and pick up a desired part from a bin of disorganized parts. An industrial-robot company recently reported success at this task using a deep neural network with eight hours of training. A drone company recently described a deep neural network that autonomously flies drones in complex real-world environments. Why are these advances happening now? For these networks to learn effectively, they require large training sets, often with millions of examples. This, combined with the large size of the networks, means that they also require large amounts of computational power. These systems are having a big impact now because the Web is a source of large training sets, and modern computers with graphics co-processors have the power to train them.

Where is this going? Expect these networks to soon take on every conceivable application. Several recent university courses on deep learning have posted their students' class projects. In just a few months, hundreds of students were able to use these technologies to solve a wide variety of problems that would have been regarded as major research programs a decade ago. We are in a kind of Cambrian explosion of these networks right now. Groups all over the world are experimenting with different sizes, structures, and training techniques, and other groups are building hardware to make them more efficient.

All of this is exciting but it also means that artificial intelligence is likely to soon have a much bigger impact on our society. We must work to ensure that these systems have a beneficial effect—and to create social structures that help integrate the new technologies. Many of the contest-winning networks are "feedforward" from input to output. These typically perform classification or evaluation of their inputs and don't invent or create anything. More recent networks are "recurrent nets," which can be trained by "reinforcement learning" to take actions to best achieve rewards. This kind of system is better able to discover surprising or unexpected ways of achieving a goal. The next generation of network will create world models and do detailed reasoning to choose optimal actions. That class of system must be designed very carefully to avoid unexpected undesirable behaviors. We must carefully choose the goals we ask these systems to optimize. If we can develop the scientific understanding and social will to guide these developments in a beneficial direction, the future is bright indeed!

SEEING OUR CYBORG SELVES

THOMAS A. BASS

Professor of English and journalism, University at Albany, SUNY;
author, *The Spy Who Loved Us*

We are still rolling down the track created by Moore's Law,
which means that news about science and technology will continue to focus on computers getting smaller, smarter, faster, and
increasingly integrated into the fabric of our everyday lives—in
fact, integrated into our bodies as prosthetic organs and eyes.
Our cyborg selves are being created out of advances not only in
computers but also in computer peripherals. This is the technology that allows computers to hear, touch, and see.

Computers are becoming better at "seeing" because of advances in optics and lenses. Manufactured lenses, in some ways
better than human lenses, are getting cheap enough to put everywhere. This is why the news is filled with stories about self-
driving cars, drones, and other technology that relies on having
lots of cameras integrated into objects.

This is also why we live in the age of selfies and surveillance.
We turn lenses on ourselves as readily as the world turns lenses
on us. If once we had a private self, this self has disappeared
into curated images of ourselves doing stuff that provokes envy
in the hearts of our less successful "friends." If once we walked
down the street with our gaze turned outward on the world,
now we walk with our eyes focused on the screens that mediate
this world. At the same time, we are tracked by cameras that

record our motion through public space, which has become monitored space.

Lenses molded from polymers cost pennies to manufacture, and the software required to analyze images is getting increasingly smart and ubiquitous. Lenses advanced enough for microscopy now cost less than a dollar. A recent issue of *Nature Photonics*, reporting on work done by researchers in Edinburgh, described cameras that use photons for taking pictures around corners and in other places the human eye can't see. This is why our self-driving cars will soon have lower insurance rates than the vehicles we currently navigate around town.

The language of sight is the language of life. We get the big picture. We focus on a problem. We see—or fail to see—each other's point of view. We have many ways of looking, and more are being created every day. With computers getting better at seeing, we need to keep pace with understanding what we're looking at.

THE REJECTION OF SCIENCE ITSELF

DOUGLAS RUSHKOFF

Media analyst; documentarian; author, *Throwing Rocks at the Google Bus*

I'm most interested by the news that an increasing number of people are rejecting science altogether. With 31 percent of Americans believing that human beings have existed in their current form since the beginning and only 35 percent agreeing that evolution happened through natural processes, it's no wonder that parents reject immunization for their children and voters support candidates who value fervor over fact.

To be sure, science has brought some of this on itself, by refusing to admit the possibility of any essence to existence and by too often aligning with corporate efforts to profit from discoveries with little concern for the long-term effects on human well-being.

But the dangers of an antiscientific perspective, held so widely, are particularly perilous at this moment in technological history. We are fast acquiring the tools of creation formerly relegated to deities. From digital and genetic programming to robots and nanotechnology, we are developing things that, once created, will continue to act on their own. They will adapt, defend, and replicate, much as life itself. We have evolved into the closest things to gods this world has ever known, yet most of us have yet to acknowledge the actual processes that got us to this point.

That so many trade scientific reality for provably false fantasy at precisely the moment when we have gained such powers may not be entirely coincidental. But if these abilities are seen as something other than the fruits of science, and are applied with utter disregard to their scientific context, I fear we will lack the humility required to employ them responsibly.

The big science story of the century—one that may even decide our fate—will be whether or not we accept science at all.

RE-THINKING ARTIFICIAL INTELLIGENCE

RODNEY A. BROOKS

Roboticist; Panasonic Professor of Robotics, emeritus, MIT; author, *Flesh and Machines*

This past year there has been an endless supply of news stories, as distinct from news itself, about artificial intelligence. Many of these stories concerned the opinions of eminent scientists and engineers who do not work in the field, about the almost immediate dangers of superintelligent systems waking up and not sharing human ethics and being disastrous for humankind. Others have quoted people in the field on the immorality of having AI systems make tactical military decisions. Still others report that various car manufacturers predict the imminence of self-driving cars on our roads. Yet others cite philosophers (amateur and otherwise) on how such vehicles will have to make life-or-death decisions.

My own opinions on these topics are counter to the popular narrative; mostly I think people are getting way ahead of themselves. Arthur C. Clarke's third law is that any sufficiently advanced technology is indistinguishable from magic. These news stories, and the experts prompting them, are jumping so far ahead of the state of the art in AI that they talk about a magic future variety of it, and once magic is involved, any consequence one desires or fears can be derived.

There has also been recent legitimate news on artificial intelligence, most of it centering on the stunning performance of deep-learning algorithms—the back-propagation ideas of the mid-1980s now extended, by better mathematics, to many more than just three network layers, and extended in computational resources by the massive computer clouds maintained by West Coast U.S. tech titans and also by the clever use of GPUs (Graphical Processing Units) within those clouds.

The most practical immediate effect of deep learning is that speech-understanding systems are noticeably better than just two or three years ago, enabling new services on the Web or on our smartphones and home devices. We can easily talk to those devices now and have them understand us. The frustrating speech interfaces of five years ago are gone.

The success of deep learning has, I believe, led many people to wrong conclusions. When someone displays a particular performance in some task—translating text from a foreign language, say—we have an intuitive understanding of how to generalize to what sort of competence the person has. For instance, we know that the person understands that language and can answer questions about which of the people in a story about a child dying in a terrorist attack, say, would mourn for months and who would feel they had achieved their goals. But the translation program likely has no such depth of understanding. One cannot apply the normal generalization from performance to competence to make similar generalizations for AI programs.

By now we have started to see a trickle of news stories running counter to the narrative of artificial intelligence's runaway success. I welcome these stories, as they strike me as bringing reality back to the debates about our future relationship to AI. There are two sorts of such stories:

The first is about the science, with many researchers now

declaring that a lot more science needs to be done to come up with learning algorithms mimicking the broad capabilities of humans and animals. Deep learning, by itself, won't solve many of the learning problems for general AI—for instance, where spatial or deductive reasoning is involved. Further, all the breakthrough results we've seen have been years in the making, and there's no reason to expect a sudden and sustained series of them, despite the enthusiasm of young researchers who weren't around during the last three waves of such predictions, in the 1950s, 1960s, and 1980s.

The second class of stories is about how self-driving cars and drivers of other cars interact. When large physical kinetic masses are in close proximity to human beings, the rate of adoption has been much slower than that of, say, Java Script in Web browsers. There has been a naïve enthusiasm that fully self-driving cars will soon be deployed on public roads. The reality is that there will be fatal accidents (even things built by incredibly smart people sometimes blow up), which will cause irrational levels of caution (given the daily death toll worldwide of more than 3,000 automobile fatalities caused by people). The latest news stories document the high accident rate of self-driving cars under test; so far, all are minor accidents and attributable to errors on the part of the other driver, the human. The cars themselves are driving perfectly, goes the narrative, and not breaking the law like all humans do, so it's the humans that are at fault. When you're arguing that those pesky humans just don't get a technology, you've already lost the argument. A lot more work must be done before self-driving cars are loosed in environments where ordinary people are also driving, no matter how shiny the technology seems to the engineers building it.

The hype in the news about AI is finally being met with a little pushback. There will be screams of indignation from true believers, but eventually this bubble will fade. We'll gradually see more and more effective uses of AI in our lives, but it will be slow and steady—not explosive and not existentially dangerous.

I, FOR ONE

JOSHUA BONGARD

Associate professor of computer science, University of Vermont;
author, *How the Body Shapes the Way We Think*

"Welcome, our new robot overlords," I will say when they arrive. As I sit here nursing a coffee, watching the snow fall outside, I daydream about the coming robot revolution. The number of news articles about robotics and AI are growing exponentially, indicating that superintelligent machines will arise shortly. Perhaps in 2017.

As a roboticist myself, I hope to contribute to this phase change in the history of life on Earth. The human species has recently painted itself into a corner and—global climate conferences and nuclear nonproliferation treaties notwithstanding—seems unlikely to find a way out with biological smarts alone: We're going to need help. And the growing number of known Earth-like yet silent planets indicates that we can't rely on alien help anytime soon. We're going to need homegrown help. Machine help. There is much that superintelligent machines could help us with.

Very, very slowly, some individuals in some human societies have been enlarging their circles of empathy: human rights, animal cruelty, and microaggressions are recent inventions. Taken together, they indicate that we are increasingly able to place ourselves in others' shoes. We can feel what it would be like to be the target of hostility or violence. Perhaps machines will help us widen these circles. My intelligent frying pan may suggest sautéed veggies over the bloody steak I'm about to drop into it. A smartphone might detect cyberbullying in a photo I'm about to upload and suggest that I think about how that might make the person in the photo feel. Better yet, we could imbue machines with the goal of self-preservation, mirror neurons to mentally simulate how others' actions may endanger their own continued existence, and the ability to invert those thought processes so that they can realize how their own actions threaten the existence of others. Such machines would then develop empathy. Driven by sympathy, they would feel compelled to teach us how to strengthen our own abilities in that regard. In short: future machines may empathize about humans' limited powers of empathy.

The same neural machinery that enables us (if we so choose) to imagine the emotional or physical pain suffered by another also allows us to predict how our current choices will influence our future selves. This is known as prospection. But humans are also lazy; we make choices now that we come to regret later. Machines could help us here, too. Imagine neural implants that can directly stimulate the pain and pleasure centers of the brain. Such a device could make you feel sick before your first bite into that bacon cheeseburger rather than after you've finished it. A passive-aggressive comment to a colleague or loved one would result in an immediate fillip to the inside of the skull.

In the same way that machines could help us maximize our powers of empathy and prospection, they could also help us minimize our agency-attribution tendencies. If you're a furry little creature running through the forest and you see a leaf shaking near your path, it's safer to attribute agency to the leaf's motion than to not: Better to believe there's a predator hiding behind the leaf than to attribute its motion to wind. Such paranoia stands you in good Darwinian stead, in contrast to another creature who thinks "Wind" and ends up eaten. It is possible that such paranoid creatures evolved into religious humans who saw imaginary predators (i.e., gods) behind every thunderstorm and stubbed toe. But religion leads to religious wars and leaders who announce, "God made me do it." Such defenses don't hold up well in modern humanist societies. Perhaps machines could help us correctly interpret the causes of each and every sling and arrow of outrageous fortune we experience in our daily lives. Did I miss my bus because I'm being punished for the fact that I didn't call my sister yesterday? My Web-enabled glasses immediately flick on to show me that bus schedules have become more erratic due to this year's cut in my city's public transportation budget. I relax as I start walking to the subway: It's not my fault.

What a wonderful world it could be! But how to get there? How would the machines teach empathy, prospection, and correct agency attribution? Most likely, they would overhaul our education system. The traditional classroom setting would finally be demolished so that humans could be taught solely in the school of hard knocks: Machines would engineer everyday situations (both positive and negative) from which we would draw the right conclusions. But this would take a lot of time and effort. Perhaps the machines would realize that rather than expose every human to every valuable life lesson, they could distill down a few important ones into videos or even text: *The plight of the underdog. She who bullies is eventually bullied herself. There's just us. Do to others what. . . .* Perhaps these videos and texts could be turned into stories, rather than delivered as dry treatises on morality. Perhaps they could be broken into small bite-sized chunks, provided on a daily basis. Perhaps instead of hypothetical scenarios, life lessons could be drawn from real plights suffered by real people and animals. Perhaps they could be broadcast at a particular time—say, 6:00 pm and 11:00 pm—on particular television channels, or whatever the future equivalent venue is.

The stories would have to be changed daily to keep things "new." And there should be many of them, drawn from all cultures, all walks of life, all kinds of people and animals, told from all kinds of angles to help different people empathize, prospect, and impute causes to effects at their own pace and in their own way. So, not "new" then, but "news."

DATA SETS OVER ALGORITHMS

ALEXANDER WISSNER-GROSS

Inventor; entrepreneur; president and chief scientist, Gemedy

Perhaps the most important news of our day is that data sets, not algorithms, might be the key limiting factor in the development of human-level artificial intelligence.

At the dawn of the field of artificial intelligence, in 1967, two of its founders famously anticipated that solving the problem of computer vision would take only a summer. Now, almost a half century later, machine-learning software finally appears poised to achieve human-level performance on vision tasks and a variety of other grand challenges. What took the AI revolution so long?

A review of the timing of the most publicized AI advances over the past thirty years suggests a provocative explanation: Perhaps many major AI breakthroughs have been constrained by a relative lack of high-quality training data sets and not of algorithmic advances. For example, in 1994, the achievement of human-level spontaneous speech recognition relied on a variant of a hidden Markov model algorithm published ten years earlier, but used a data set of spoken *Wall Street Journal* articles and other texts available only three years earlier. In 1997, when IBM's Deep Blue defeated Garry Kasparov to become the world's top chess player, its core NegaScout planning algorithm was fourteen years old, whereas its key data set of 700,000 Grandmaster chess games (known as the "The Extended Book") was only six years old. In 2005, Google software achieved breakthrough performance at Arabic- and Chinese-to-English translation based on a variant of a statistical machine-translation algorithm published seventeen years earlier, but used a data set with more than 1.8 trillion tokens from Google Web and News pages gathered the same year. In 2011, IBM's Watson became the world *Jeopardy!* champion using a variant of the mixture-of-experts algorithm published twenty years earlier, but utilized a data set of 8.6 million documents from Wikipedia, Wiktionary, Wikiquote, and Project Gutenberg updated the year before. In 2014, Google's GoogLeNet software achieved near-human performance at object classification using a variant of the convolutional neural network algorithm proposed twenty-five years earlier, but was trained on the ImageNet corpus of approximately 1.5 million labeled images and 1,000 object categories first made available only four years earlier. Finally, in 2015, Google DeepMind announced that its software had achieved human parity in playing twenty-nine Atari games by learning general control from video, using a variant of the Q-learning algorithm published twenty-three years earlier, but the variant was trained on the Arcade Learning Environment data set of over fifty Atari games made available only two years earlier.

Examining these advances collectively, the average elapsed time between key algorithm proposals and corresponding advances was about eighteen years, whereas the average elapsed time between key data-set availabilities and corresponding advances was less than three years, or about six times faster, suggesting that data sets might have been limiting factors in the advances. In particular, one might hypothesize that the key algorithms underlying AI breakthroughs are often latent, simply needing to be mined out of the existing literature by large, high-quality data sets and then optimized for the available hardware of the day. Certainly, in a tragedy of the research commons, attention, funding, and career advancement have historically been associated more with algorithmic than data-set advances.

If correct, this hypothesis would have foundational implications for future progress in AI. Most important, the prioritized cultivation of high-quality training data sets might allow an order-of-magnitude speed-up in AI breakthroughs over purely algorithmic advances. For example, we might already have the algorithms and hardware that will enable machines in a few years to write human-level long-form creative compositions, complete standardized human examinations, or even pass the Turing Test, if only we trained them with the right writing, examination, and conversational data sets. Additionally, the nascent problem of ensuring AI friendliness might be addressed by focusing on data-set rather than algorithmic friendliness—a potentially simpler approach.

Although new algorithms receive much of the public credit for ending the last AI winter, the real news might be that prioritizing the cultivation of new data sets and research communities around them could be essential to extending the present AI summer.

BIOLOGICAL MODELS OF MENTAL ILLNESS REFLECT ESSENTIALIST BIASES

BRUCE HOOD

Professor, School of Experimental Psychology, University of Bristol; author, *The Domesticated Brain*

In 2010, in England alone, it was estimated that mental illness cost over £100 billion to the economy. Around the same time, the cost in the U.S. was estimated at $318 billion annually. It is important that we do what we can to reduce this burden. However, we have mostly been going about it the wrong way, because the predominant models of mental illness do not work. They are mostly based on the assumption that there are discrete underlying causes, but this approach to mental illness reflects an essentialist bias we readily apply when trying to understand complexity.

Humans are, of course, complex biological systems, and the way we operate requires sophisticated interactions at many levels. Remarkably, it has taken more than a century of research and effort to recognize that when things break down, they involve multiple systems of failure; yet, until the last couple of years, many practitioners in the West's psychiatric industry have been reluctant to abandon the notion that there are qualitatively distinct mental disorders with core causal dysfunctions. Or at least that's how the treatment regimes seem to have been applied.

Ever since Emil Kraepelin, at the end of the 19th century, advocated categorizing mental illnesses into distinct disorders with specific biological causes, research and treatment has focused on building classification systems of symptoms as a way of mapping the terrain for discovering root biological problems and corresponding courses of action. This medical-model approach led to development of clinical nosology and accompanying diagnostic manuals such as the *Diagnostic and Statistical Manual of Mental Disorder* (DSM), whose fifth edition was published in 2013. However, that year the National Institute of Mental Health announced that it would no longer be funding research projects that relied solely on DSM criteria. This is because the medical model lacks validity.

A recent analysis by Denny Borsboom in the Netherlands revealed that 50 percent of the symptoms of the DSM are correlated, indicating that co-morbidity is the rule, not the exception, which explains why attempts to find biological markers for mental illness either through genetics or imaging have proved largely fruitless. It does not matter how much better we build our scanners or refine our genetic profiling: Mental illness will not be reducible to Kraepelin's vision. Rather, new approaches consider symptoms as sharing causal effects rather than arising from an underlying primary latent variable.

It's not clear what will happen to the DSM, as there are vested financial interests in maintaining the medical model, but in Europe there is a notable shift toward symptom-based approaches to treatment. It is also not in our nature to consider human complexity other than with essentialist biases. We do this for race, age, gender, political persuasion, intelligence, humor, and just about every dimension we use to describe someone—as if these attributes were at the core of who they are.

The tendency of the human mind is to categorize the world—to carve nature up at its joints, as it were. But in reality, experience is continuous. The boundaries we create are more for our benefit than a reflection of any true existing structures. As complex biological systems, we evolved to navigate the complex world around us and thus developed an ability to represent it in the most useful way, as discrete categories. This is a fundamental feature of our nervous system, from the input of raw sensory signals to the output of behavior and cognition. Forcing nature into discrete categories optimizes the processing demands and the number of needed responses, so it makes sense from an engineering perspective. The essentialist perspective continues to shape the way we go about building theories to investigate the world. Maybe it's the best strategy when dealing with unknown terrain—assume patterns and discontinuities with broad strokes before refining your models to reflect complexity. The danger lies in assuming that the frameworks you construct are real.

NEUROPREDICTION

ABIGAIL MARSH

Associate professor of psychology, Georgetown University

The Cartesian wall between mind and brain has fallen. Its disintegration has been aided by the emergence of a wealth of new techniques in collecting and analyzing neurobiological data, including neuroprediction, which is the use of human-brain imaging data to predict how the brain's owner will feel or behave in the future. The reality of neuroprediction requires accepting the fact that human thoughts and choices are a reflection of basic biological processes. It has the potential to transform fields like mental health and criminal justice.

In mental health, advances in identifying and treating psychopathology have been limited by existing diagnostic practices. In other fields of medicine, new diagnostic techniques such as genetic sequencing have led to targeted treatments for tumors and pathogens and major improvements in patient outcomes. But mental disorders are still diagnosed as they have been for 100 years—using a checklist of symptoms derived from a patient's subjective reports or a clinician's subjective observations. This is like trying to determine whether someone has leukemia or the flu based on subjective evaluations of weakness, fatigue, and fever. The checklist approach not only makes it difficult for a mental-health practitioner to determine what afflicts a patient—particularly if the patient is unwilling or unable to report his symptoms—but also provides no information about what therapeutic approach will be most effective.

In criminal justice, parallel problems persist, in sentencing and probation. Making appropriate sentencing and probation decisions is hampered by the difficulty of determining whether a given offender is likely to re-offend after being released. Such decisions, too, are based on largely subjective criteria. Some who likely would not recidivate are often detained for too long, and some who will are released.

Neuroprediction may yield solutions to these problems. One recent study found that the relative efficacy of different treatments for depression could be predicted from a brain scan measuring metabolic activity in the insula. Another found that predictions about whether paroled offenders would recidivate were improved using a brain scan that measured hemodynamic activity in the anterior cingulate cortex. Neither approach is ready for widespread use yet, in part because predictive accuracy at the individual level is still only moderate, but inevitably they—or improvements on them—will be.

This would be an enormous advance in mental health. Currently, treatment outcomes for disorders like depression remain poor; up to 40 percent of depressed patients fail to respond to the first-line treatment, the selection of which still relies more or less on guesswork. Using neuroprediction to improve this statistic could dramatically reduce suffering. Because brain scans are expensive and their availability limited, however, there would be disparities in access.

Neuroprediction of crime presents a different scenario, as its primary purpose would be to improve outcomes for society (less crime, fewer resources spent on needless detentions) rather than for the potential offender. It's hard to imagine this becoming accepted practice without a shift in our focus away from retribution and toward rehabilitation. In furthering understanding of the biological basis of persistent offending, neuroprediction may help in this regard. Regardless, neuroprediction, at least the beta version of it, is here. Now is the time to consider how to harness its potential.

THE THIN LINE BETWEEN MENTAL ILLNESS AND MENTAL HEALTH

JOEL GOLD

Psychiatrist; clinical associate professor of psychiatry, NYU School of Medicine; co-author (with Ian Gold), *Suspicious Minds*

It is discomfiting for many of us to contemplate this fact. We prefer to imagine a nice thick wall between us, the "Well," and them, the "Mad."

In one episode of *The Simpsons*, Homer is psychiatrically hospitalized by mistake. His hand is stamped "Insane." When his psychiatrists come to believe he's not and release him, they stamp his hand "Not Insane." But sanity is not binary; it's a spectrum on which we all lie. Overt madness might be hard to miss, but what is its opposite? There is clear evidence that large numbers of people who have no psychiatric diagnosis and are not in need of psychiatric treatment experience symptoms of psychosis, notably hallucinations and delusions. A recent study in *JAMA Psychiatry* surveyed more than 30,000 adults from nineteen countries and found that 5 percent had heard voices at least once in their life. Most of these people never developed full-blown psychosis of the type observed in a person with, say, schizophrenia. An older study reported that 17 percent of the general nonclinical population had experienced psychosis at some point.

It gets even more slippery: It isn't always clear if an experience is psychotic or not. Why is someone who believes that the U.S. government is aware of alien abductions deemed not delusional but merely a conspiracy theorist, yet someone who believes that he himself has been abducted by aliens is likely considered delusional?

The psychosis continuum has important clinical ramifications. Unfortunately, that is news to many mental-health practitioners. It's easy to see the neurobiological parallels between antidepressant medication improving mood, anxiolytic medication reducing panic, and antipsychotic medication ameliorating hallucinations. But ask a psychiatrist about providing psychotherapy to people suffering from these symptoms and, again, the wall comes up. At least here in New York City, many people with depression and anxiety seek relief in therapy. Very few of those with psychosis are afforded its benefits, despite the fact that therapy works in treating psychotic symptoms. And here is where the lede has been buried.

Cognitive behavioral therapy (CBT)—one of the most practiced forms of therapy—while commonly applied to mood, anxiety, and a host of other psychiatric disorders, also works with psychosis. This might seem inherently contradictory. By definition, a delusion is held despite evidence to the contrary. You aren't supposed to be able to talk someone out of a delusion. If you could, it wouldn't be a delusion, right? Surprisingly, this is not the case.

And here we return to our thin line. Early in CBTp, the therapist "normalizes" the psychotic experiences of the patient—perhaps going so far as to offer his own strange experiences—thereby reducing stigma and forging a strong therapeutic bond with the patient, who is encouraged to see himself not as "less than" his doctor but further along the spectrum (the continuum model). The patient is then educated as to how stressors like child abuse or cannabis use can interact with preexisting genetic risk factors and is encouraged to reflect on the effects his life experiences might have on his symptoms (the vulnerability-stress model). Finally, the therapist reviews an Activating event, the patient's Belief about that event, and the Consequences of holding that belief (the ABC model). Over time, the clinician gently challenges it, and ultimately patient and doctor together reevaluate the belief. CBTp can be applied to hallucinations as well as to delusions.

CBTp has about the same therapeutic benefit as the older antipsychotic medication chlorpromazine (Thorazine) and the newer antipsychotic olanzapine (Zyprexa). This does not mean, of course, that people shouldn't take antipsychotic medication when appropriate; they certainly should. The reality, however, is that many do not, and it's not hard to understand why. These medications, while often life-saving (for the record, I have prescribed antipsychotics thousands of times), often have adverse effects. Impaired insight (the ability to reflect on one's inner experiences and to recognize that one is ill) is also a significant impediment to medication adherence.

Here, CBTp can have several ancillary benefits. First, the therapy can improve insight and thereby adherence. Second, if a patient refuses to take medication but is willing to engage in CBTp, she will do better than with no treatment at all. Finally, people receiving CBTp might ultimately require lower doses of antipsychotic medication, diminishing its toxicity and, again, increasing adherence.

The utility of CBTp shouldn't be news, as evidence of its efficacy has been replicated over and again, but it remains so, sadly even in the mental health community, especially in the United States. While CBTp is a first-line treatment for psychosis in the U.K., you would be hard-pressed to find a U.S. psychiatrist who could describe how it's practiced. Good luck finding a mental-health practitioner trained to do it. But the good news is that the news is spreading. However slowly, more clinicians are becoming aware, being trained, and practicing CBTp. More practitioners will become available to more patients, who will then receive better care (optimally, along with other well-established psychosocial interventions, like family therapy and supported employment), and we will see improved medical outcomes.

If this news sticks—and I think it will—it will have a great humanizing effect in the way society views people suffering from psychosis. After all, while there are psychotic aspects in all of our minds, it is assuredly just as true that there are healthy parts of even the most stricken.

THEODIVERSITY

ARA NORENZAYAN

Professor of psychology, University of British Columbia; author, *Big Gods: How Religion Transformed Cooperation and Conflict*

Theodiversity is to religion what biodiversity is to life. There are, by some accounts, more than 10,000 religious traditions in the world. Every day, somewhere in the world, a new religious movement is in the making. But this theodiversity—a term I borrow from Toby Lester—is not evenly distributed in human populations, any more than biodiversity is evenly distributed on the planet.

The overwhelming majority of religious movements throughout history are failed social experiments. Most never take hold; those that do, don't last for long, and of those that last for a while, most stay small. Then there are the "world religions." Christianity, Hinduism, and Islam especially, have been growing at a brisk pace. Buddhism is much smaller and not growing much but is still a sizable presence on the world stage. We are at a point in time when just a few religious traditions have gone global, making up the vast majority of believers in the world.

This fact is detailed in a landmark Pew Research Center report, released on April 2, 2015. It's the most comprehensive and empirically derived set of projections based on data, age, fertility, mortality, migration, and religious conversion/de-conversion for multiple religious groups around the world. Barring unforeseen shocks, if current demographic and social trends keep up, by 2050:

- Possibly for the first time in history, there will be as many Muslims as Christians in the world. Together, these two faiths will represent more than 60 percent of the world's projected population of 9.5 billion.
- 40 percent of the Christians will live in sub-Saharan Africa (which will have the largest share of Christians), compared with 15 percent in Europe—so the epicenter of Christianity will have finally shifted from Europe to Africa.
- India, while maintaining a Hindu majority, will have the largest Muslim population in the world, surpassing Indonesia and Pakistan.
- All the folk religions of the world combined will comprise less than 5 percent of the world's population.
- 1.3 billion people, or 13.5 percent of the world's population in 2050, will be non-religious.

One might think that religious denominations that have most successfully adapted to secular modernity are the ones thriving the most. But the evidence gleaned from the Pew report and other studies points in the opposite direction. Moderate denominations are falling behind in the cultural marketplace. They are the losers, caught between secular modernity and the fundamentalist strains of the major world religions—strains that are gaining steam as a result of conversion, higher fertility rates, or both.

There are different types and intensities of disbelief. That's why the non-religious are another big ingredient of the world's dynamically changing theodiversity. Combined, they would be the fourth largest "world religion." There are the atheists, but many nonbelievers are instead apatheists—indifferent toward and not opposed to religion. And there is a rising demographic tide of people who see themselves as "spiritual but not religious." This do-it-yourself, custom-made spirituality is filling the void that the retreat of organized religion has created in the secularizing countries. You can find it in yoga studios, meditation centers, the holistic health movement, and eco-spirituality.

Theodiversity once was the exclusive subject matter of the humanities. But it is now a focal point of a budding science/humanities collaboration. The religious diversification of humankind in historical time poses fascinating questions and challenges for the new science of cultural evolution. These are times of renewed anxiety about (real or imagined) cultural conflict between religions and conflict between religions and secular modernity. Quantifiable, evidence-based, and nuanced understanding of the complexities of theodiversity is important now more than ever.

MODERNITY IS WINNING

GREGORY PAUL

Independent researcher; author, *The Princeton Field Guide to Dinosaurs*

Having long been interested in the probability that cyberintelligence will soon replace humanity, I could cite frequent news coverage of efforts to produce advanced artificial intelligence as the most important news. But that's an obvious subject, so I won't.

Instead, I will discuss a much more obscure science-news item of potentially great long-term import. It's how some privately funded, commercial fusion-power projects are being initiated. The intent is to produce the unlimited cheap power that government-backed projects have failed to deliver.

What few recognize is how the fusion news is tied to a much more prominent story. A lately released major Pew analysis projects a rise in theism in many developing nations in coming decades. This follows a major Pew survey showing a rapid rise of non-theism in the United States. The common opinion is that while religion is continuing to sink in the Western democracies, it's making a comeback in less stable and prosperous nations, in a historical rebuff to modernity. The resulting reactionary theism is often adopting a virulent form that afflicts the secular democracies and threatens the future of modern civilization.

What does the news on fusion power have to do with the news about reactionary religion? To see the connection, we'll start with Arthur C. Clarke.

The SciFi Channel recently presented their version of

Clarke's classic novel *Childhood's End*. Written in the early 1950s, *CE* is like many of Clarke's futurist fictional works, in which he repeatedly predicted that in the late 1900s and into the 2000s the world community would become increasingly secular, progressive, and pacific, forming a modernist planetary demi-utopia. This rested on a science-based hope. The technologist Clarke presumed that fusion energy was a readily solvable science and engineering problem, and that hydrogen-to-helium reactors would be providing the entire world with all the power we could use by the turn of the 21st century. The resulting universal prosperity would elevate everyone into at least the secure middle-class affluence that studies show result in strongly atheistic, liberal, lower-violence societies at the expense of (often dysfunctional) tribalistic religion.

That hasn't happened. Fission is easy to achieve—so easy that when uranium was more highly enriched back in the Precambrian, reactors spontaneously formed in uranium ores. Sustained fusion thermonuclear reactions so far occur only in the extreme pressure-temperature conditions at the centers of stars, and getting them to work elsewhere has proved highly difficult. Lacking fusion reactors, we have had to continue to rely mainly on fossil fuels.

Had fusion power come online decades ago, the Saudis would not have had loads of oil-generated cash to fund the virulent and widespread Wahhabist mosques and schools that have helped spread hyperviolent forms of Islam. Without cheap fusion power, much of the world remains mired in the lack of economic opportunity that breeds supernaturalistic extremism. Since the end of the Cold War reduced mass lethal violence from atheistic communists, a few million have died in war-level conflicts that share a strong religious component. Muslims are causing the most trouble. But so are Christians in sub-Saharan Africa and in Russia, where the Orthodox Church backs Putin. Even the Buddhism that Clarke saw as peaceful has gone noxious in parts of Asia, as have many Hindus in India.

But as bad as the situation is, it is not as bad as it may seem. The Pew projections are based on a set of dubious assumptions, including that the faith people are born into is the most critical factor in predicting future patterns because the pious tend to reproduce more rapidly than the secular. But trends measured by the World Values Survey indicate that religiosity is declining in most of the world. That's because conversion from theism to secularism is trumping reproduction, and that in turn is because the global middle class is on the rise, leading to retreat from religion (note that religion is not a big problem in South America or most of eastern Asia because secularism is waxing in those regions).

So why has a portion of modern religion become so venomous? In part, it's a classic counterreaction to the success of secularization. But as troublesome as they often are, such reactionary movements tend to be temporary—remember how gay-bashing was once a major sociopolitical tool of the American right? Toxic theism is a symptom of a power-hungry world.

Clarke may well have been right that fusion-power production would have helped produce a much better 21st-century world. Where he was over-optimistic was in thinking that fusion reactors would be up and running decades ago. Clarke lived long enough to be distressed when his power dream remained unfulfilled and the unpleasant social consequences became all too clear. Whether efficient hydrogen-fusing plants can be made practical in the near future is open to question; even if they can, we will have had to put up with decades of brutal strife fueled by too much religion.

That's big news. But the more important (mostly unacknowledged) news is that modernity is winning as theism retracts in the face of the prosperity made possible by modern science and technology.

RELIGIOUS MORALITY IS MOSTLY BELOW THE BELT

MICHAEL McCULLOUGH

Director, Evolution and Human Behavior Laboratory, University of Miami; author, *Beyond Revenge*

In most facets of life, people are perfectly content to let other people act in accordance with their tastes, even when those tastes differ from their own. The supertasters of the world, for instance—that 15-or-so percent of us whose tongues are so densely packed with taste buds that they find the flavors of many common foods and drinks too rich or too bitter to enjoy—have never taken to the streets to demand global bans on cabbage or coffee. And the world's normal tasters, who clearly have a numerical advantage over the supertasters, have never tried to force the supertasters into eating and drinking things they don't like.

Religion sits at the other end of the "*Vive la différence*" spectrum. The world's major religions, practiced by five of every seven people on the planet today, all teach people to concern themselves with other people's behavior—and not just the behavior of fellow believers. They often teach their adherents to take an interest in outsiders' behavior as well. Why? Recent scientific work is helping to solve this puzzle—and it has yielded a discovery that Freud would have loved.

There are two popular families of theory that seek to explain why religion causes people to praise some behaviors and condemn others. According to the first, people espouse religious beliefs—particularly a belief in an all-seeing sky god who watches human behavior and metes out rewards and punishments (in this life or the next)—because it motivates them (and others) to be more trusting, generous, and honest than they otherwise would be.

But a newer line of theorizing, called reproductive-religiosity theory, proposes that religious morality is not fundamentally about encouraging cooperation. Instead, people primarily use religion to make their social worlds more conducive to their own preferred approaches to sex, marriage, and reproduction. For most of the world's religions over the past several millennia (which have historically thrived in state societies whose primary economic driver is agricultural production), the preferred sexual strategy has involved monogamy, sexual modesty, and the stigmatization of extramarital sex (arguably because it helps to ensure paternity, thereby reducing conflict over heritable property). Reproductive-religiosity theory has a lot to commend it: In a recent cross-cultural study involving over 16,000 participants from fifty-six nations, researchers found that religious young people (from every region of the world and every conceivable religious background) were more averse to casual and promiscuous sex than were their less religious counterparts. (Tellingly, in most regions religion appeared to regulate sexuality more strongly for women than for men.)

Both theories predict that highly religious people will espouse stricter moral standards than less religious people will—and virtually every survey ever conducted supports this prediction. Religious belief seemingly influences people's views on topics as varied as government spending, immigration, social inequality, the death penalty, and euthanasia, not to mention homosexuality, same-sex marriage, abortion, pornography, and the role of women in society. But for most of the issues not explicitly related to sex, marriage, and reproduction, religion's influence appears slight. For the sex-related issues, religion's apparent influence is much stronger.

Reproductive-religiosity theory, positing as it does that religion is mostly about sex, makes an even bolder prediction: After you have statistically accounted for the fact that highly religious people have stricter sexual morals than the less religious (for instance, they are more disapproving of homosexuality, sexual infidelity, abortion, premarital sex, and women in the workplace), then they will appear not to care much more than the nonreligious about violations involving dishonesty and broken trust (transgressions such as stealing, fare dodging, tax dodging, and driving under the influence, for example).

This bolder prediction has now been supported resoundingly, and not only among Americans but also in a study involving 300,000 respondents from roughly ninety countries. Highly religious people around the world espouse stricter moral attitudes regarding both prosociality and sex, but their stern moral attitudes toward honesty-related infractions seem, from a statistical point of view, to be mostly along for the ride. It is sex, marriage, and reproduction—and not trust, honesty, and generosity—that lie at the core of moralization for most practitioners of the world religions.

As I mentioned earlier, Freud would have loved these results, but perhaps we shouldn't be too surprised that religion's most potent effects on morality relate to sex, marriage, and reproduction. After all, sex is close to the engine of natural selection, so it is not unlikely that evolution has left us highly motivated to seek out any tool we can—even rhetorical ones of the sort that religion provides—to make the world more conducive to our own approaches to love and marriage. Even so, the intimate link between religion and sexual morality is a particularly important element of certain recent geopolitical developments, so we need to understand it better than we do now.

Over the past several years, Islamic extremists in the Middle East and sub-Saharan Africa have been systematically perpetrating sexual atrocities against girls and women, and as they have done so, they have drawn explicitly on the moral support of their religious traditions. Make no mistake: War rape is nothing new, all of it is appalling, and none of it is acceptable. But to understand what is happening right now—at a time when Boko Haram fighters capture and then seek to impregnate hundreds of Nigerian schoolgirls, at a time when ISIS fighters capture thousands of Yazidi girls and women and consign them to lives of unceasing sexual terror—we need to figure out how sets of religious beliefs ordinarily bolstered to support monogamy and "family values" can transform gang rape and sexual slavery into religious obligations, not to mention the perquisites of having God on your side.

A SCIENCE OF THE CONSEQUENCES

LUCA DE BIASE

Journalist; editor, Nòva24, of Il Sole 24 Ore

Can something that didn't happen be news? Can something that didn't happen be interesting and important? To answer, one needs to add duration to the notion of news. The answer cannot be about events in a particular moment in time, but it can very well be about news that will stay news—news that has consequences.

"Big news" is news that succeeds in framing the debate, news that is often controversial. It is always interesting and only sometimes important. "News that will stay news" is different: It can be underreported but it will last for a long time. Rather than news about facts, it is a story with a lasting effect on many facts—a story that makes history. It is a narrative that guides human choices in building the future. Rarely do we find news about the emergence of a new narrative; newspapers are not made to do that. One reads that kind of emergence not in the news but between the lines of the news.

Notions such as "climate change," "gene-editing," and "nanotechnology" have branded a set of important research paths that otherwise would have appeared less interesting and may have gone unnoticed or even been misunderstood. But the convergence of science and communication is not enough to deal with the great transformation the world is facing—which needs aware audiences but, even more, informed citizens. The very notion of "science-based policy" needs improvement. A simple version of this notion has led to better-informed decisions in fields such as health and education, but it is still differently understood in political and cultural contexts, particularly those in which ideology and religion figure in the decision-making process. Global matters need a common understanding of problems and possible solutions.

The United Nations Conference on Climate Change in Paris was an example of a winning relationship between science and policy, even though it took too long to happen and achieved too little. Politicians will always be responsible for decision making, but such urgent global problems need a better sort of "science-based policy." It is a question not only of politicians listening to scientists but also of science that is effective in self-governing and developing better-informed citizens.

Gene editing provides an important case study. The U.S. National Academy of Sciences, the National Academy of Medicine, the Chinese Academy of Sciences, and the U.K.'s Royal Society recently held a summit in Washington with international experts to discuss the scientific, ethical, and governance issues associated with human gene-editing research. The idea was to call for a moratorium on using the CRISPR/Cas9 technology to edit the human genome in a permanent and heritable way, because unintended consequences were to be expected. Among those calling for the moratorium were CRISPR's inventors, but the summit ended with no big decisions. The national academies opted for a continuing discussion.

One winning argument for not making decisions was made by George Church of Harvard, who works in the field of human gene-editing: He opposed the idea of a ban by claiming it would strengthen underground research, black markets, and medical tourism—suggesting that science in a globalized economy is pretty much out of control. This is the kind of story one reads behind the lines in the news.

The debate about artificial intelligence, led by Stephen Hawking, was also about science out of control. As are discussions about robots that can take over human jobs. Science facts and news are creating a big question mark: *Is* science out of control? Could it be different? Could a science exist that was under control?

The old answer was more or less the following: Science is about finding things out; ethics or policy will serve to decide what to do about them. That kind of answer is no help anymore, because science is very much able to change how things are, while the growing demand for science-based policy enables it to take part in the decision-making process. If a scientific narrative converges with *laissez-faire* ideology and the idea of complexity, the decision-making process becomes increasingly difficult and the situation seems to go out of control.

Science needs to do something about this. Ethics informs individual decision making but needs help dealing with complexity. Policy makes collective decisions but needs theories about the way the world is changing. Science is called to take part in decision making. But how can it, without losing its soul?

There cannot be a science under control. But there can be a science that knows how to empirically deal with choices and gets better at self-government. The current "news that will stay news" may be the failure of the Paris conference scientists to decide about human gene-editing: It is a story that will stay news until an improved "science of the consequences" story comes along. Thus the scientific method must take into account the consequences of research. If the decision-making process is no longer confined to ethics and politics, epistemology is called upon to spring into action.

CREATION OF A "NO ETHNIC MAJORITY" SOCIETY

DAVID BERREBY

Journalist; author, *Us and Them*

Throughout the history of the United States, white people have been the dominant ethnic group. The exact definition of this "race" has changed over time, as successive waves of immigrants (Germans in the 18th century, Irish in the 19th, Italians and Jews in the late 19th and early 20th) worked to be included in the privileged category (as recounted, for example, in Noel Ignatiev's *How the Irish Became White*). Whatever "whiteness" meant, though, its predominance persisted—both statistically and culturally (as the no-asterisk default definition of "American"). Even today, long after the legal structure of discrimination was undone, advantages attach to white identity in the job market, housing, education, or in any encounter with authority. Not unrelatedly, life expectancy for whites is greater than for African-Americans.

But this era of white predominance is ending.

Not long after 2040, fewer than half of all Americans will identify as white, and the country will become a majority-minority nation—47 percent white, 29 percent Hispanic, 13 percent African-American, and 11 percent "other," according to U.S. Census Bureau projections. Given this demographic shift, the habits and practices of a white-dominated society cannot endure much longer. Political, legal, cultural, and even personal relations between races and ethnic groups must be renegotiated. In fact, this inevitable process has already begun. And that's news that will stay news, now and for a long time to come. It is driving a great deal of seemingly unrelated events in disparate realms, from film criticism to epidemiology.

I'll begin with the most obvious signs. In the past two years, non-whites have succeeded as never before in changing the terms of debates that once excluded or deprecated their points of view. This has changed both formal rules of conduct (for police, for students) but also unwritten norms and expectations. Millions of Americans have recently come to accept the once fringe idea that police frequently engage in unfair conduct based on race. And many now support the removal of memorials to Confederate heroes, and their flag, from public places.

Meanwhile, campuses host vigorous debates about traditions that went largely unquestioned two or three years ago. (It is now reasonable to ask why, if Princeton wouldn't name a library after Torquemada, it should honor the fiercely racist Woodrow Wilson.) The silliness of some of these new disputes (like Oberlin students complaining that the college dining hall's Chinese food was offensive to Chinese people) shouldn't obscure the significance of the trend. We are seeing inevitable ethnic renegotiation, as what was once "harmless fun" (like naming your football team the Redskins) is redefined as something no decent American should condone.

It's nice to imagine this political and cultural reconfiguring as a gentle and only slightly awkward conversation. But evidence suggests that the transition will be painful and its outcome uncertain.

Ethnic identity (like religious identity, with which it is often entangled) is easy to modify over time but difficult to abandon. This is especially true when people believe their numbers and influence are declining. In that situation, they become both more aware of their ethnicity and more hostile to "outsiders." (In a 2014 paper, for example, the social psychologists Maureen A. Craig and Jennifer A. Richeson found that white citizens who'd read about U.S. demographics in 2042 were more likely to agree with statements like "It would bother me if my child married someone from a different ethnic background" compared with whites who had read about 2010's white-majority demographics.*) Such feelings can feed a narrative of lost advantage even when no advantage has been lost. Though whites remain privileged members of American society, they can experience others' gains toward equality as a loss for "our side." The distress of white people over the loss of their predominance—a sense that "the way things were before was better"—has rewarded frankly xenophobic rhetoric and the candidates who use it.

We should not imagine, though, that this distress among some whites is manifested merely rhetorically. In a recent analysis of statistics on sickness and death rates, the economists Anne Case and Angus Deaton found that middle-aged white people in the United States have been dying by suicide, drug abuse, and alcohol-related causes at extraordinary rates.† The historian and journalist Josh Marshall has pointed out that this effect is strongest among the people who, lacking other advantages, had the most stake in white identity: less educated, less skilled, less affluent workers. (Other scholars have disputed details of Case and Deaton's analysis but not its overall point.)

If the Case-Deaton statistics reflected only economic distress, then middle-aged working-class people of other ethnic groups

---

* http://pss.sagepub.com/content/25/6/1189

† http://www.nber.org/papers/w21279

should also be missing out on the general health improvements of the last few decades. This is not so. Unskilled middle-aged African-Americans, for example, have lower life expectancy than equivalent whites. Yet their health measures continually improved over the period during which those of whites stalled.

For this reason, I think Marshall is right and the Case-Deaton findings signal a particularly racial distress. The mortality rates correlate with loss of privilege, of unspoken predominance, of a once undoubted sense that "the world is ours." Over the next ten or twenty years, this ongoing news could turn into a grim story of inter-ethnic conflict.

Can scientists and other intellectuals do anything to help prevent the inevitable ethnic reconfiguration from being interpreted as a zero-sum conflict? I think they can. For one thing, much is unknown about the psychology, and even physiology, of loss of ethnic advantage. Much could be learned by systematic comparative research on societies in which relations among social groups were swiftly renegotiated so that one group lost privilege. South Africa after the fall of apartheid is one such place; Eastern Europe during and after the fall of Communism may be another.

We could also sharpen up our collective understanding of the slippery psychology of ethnic threat with an eye toward finding methods to understand and cope with such feelings. To do that, we need to take people's perceptions about identity seriously. Happy talk about the wonders of diversity and the arc of history bending toward justice will not suffice. We need to understand how, why, and when some people on this inevitable journey will experience it as a loss.

INTERCONNECTEDNESS

IRENE PEPPERBERG

Research associate, lecturer, Harvard University; adjunct associate professor, Brandeis University; author, *Alex & Me*

> No man is an island
> Entire of itself . . .

John Donne wrote these words almost 400 years ago, and (aside from the sexism of the male noun) his words are as true now as they were then. I believe they will be just as true in the future, and apply to scientific discovery as well as to philosophy. The interconnectedness of humans, and of humans and their environment, that science is demonstrating today is just the beginning of what we will discover and is the news very likely to be discussed in the future.

From the science of economics to that of biology, we are learning how the actions and decisions of each of us affect the lives of all others. The coal-fired power plants of India, China, and elsewhere affect the climate of us all, as does the ongoing deforestation of the Amazon. The nuclear disaster in Japan shaped how we view one alternative energy source. But we now know that our health (particularly our microbiome) is affected not only by what we put into our mouths but, somewhat surprisingly, also by the company we keep. Recent studies show that decisions about the removal of an invasive species affect its entire surrounding ecological web as much as decisions concerning the protection of an endangered species.

One need not necessarily buy into Donne's somewhat dark worldview to appreciate the importance of his words. Interconnectedness means that the scientists of the world work to find a cure for a disease such as Ebola that has, so far, primarily been limited to a few countries. It also means that governments recognize how reacting to the plight of refugees from wartorn areas halfway around the globe could be a means of enriching rather than impoverishing one's country.

Whether we look at social media, global travel, or any other form of interconnectedness, news of its importance is here to stay.

EARLY LIFE ADVERSITY AND COLLECTIVE OUTCOMES

LINDA WILBRECHT

Associate professor, UC Berkeley Department of Psychology and
Helen Wills Neuroscience Institute

When fear and tension rise across racial and ethnic divisions, as they have in recent years, genetic arguments to explain behavioral differences can quickly become popular. However, we know racial and ethnic groups may also be exposed to vastly different experiences likely to strongly affect behavior. Despite our seemingly inexhaustible interest in the nature/nurture debate, we are only starting to learn how the interaction of genes with experience may alter the potential of individuals, and to see how individual decision-making styles can alter the potential wealth of nations.

A recent captivating news image depicted two sets of male identical twins, mixed up as infants and raised in separate families in Colombia. The boys and their families assumed they were fraternal twins, who share genes only as much as siblings do and therefore don't look alike. Only in adulthood did the young men discover the mistake and find their identical twin brothers through the recognition of friends. One mixed pair of twins grew up in the city and the other in the countryside with far more modest resources. We would like to know how these different environments altered these men's personalities, preferences, intelligence, and decision making, when their genes were the same. We're all probably comfortable with the idea that trauma, hardship, or parenting style can affect our emotional development and emotional patterns even in adulthood. But it's less clear how early experience might affect how we think and make decisions. Is one identical twin, because he was raised in a different situation, more likely to save money, repeat a mistake, take a shortcut, buy lottery tickets, resist changing his mind? Or would the identical-twin pair make the same choices regardless of upbringing? The answers from these twins are still emerging, and the sample is, of course, anecdotally small. If we knew the answers, it might change how we view parenting and investment in child care and education.

A growing body of work now effectively models this "twins raised apart" situation in genetically identical strains of inbred mice or genetically similar rats. Rodents get us away from our cultural biases and can be raised in conditions that model human experiences of adversity and scarcity in infancy and childhood. In one early-life stress model, the mother rodent is not given adequate nesting material and moves about the cage restlessly, presumably in search of more bedding. In other models, the rodent pups are separated from the mother for parts of the day or are housed alone after weaning. These offspring are then compared to offspring that have been housed with adequate nesting material and have not been separated more than briefly from their mother or siblings.

Investigators first focused early-life-adversity research in rodent models on emotional behavior. This research found that early-life adversity increased adult stress and anxious behavior. More recent studies, including some from my lab, find that early-life adversity can also affect how rodents think—how they solve problems and make decisions. Rats and mice subjected to greater early-life stress tend to be less cognitively flexible (stubbornly applying old rules in a laboratory task after new rules are introduced), and they may also be more repetitive or forgetful. Some of the differences in behavior have faded as the animals age, but others grow stronger and persist into adulthood. I hesitate to say that one group is smarter, since it's hard to determine what would be optimal for a wild rodent. We might see stubborn, inflexible, or repetitive behavior as unintelligent in a laboratory test. In some real-world situations, the same behavior might be admirable, as "grit" or perseverance.

Competing theories attempt to explain these changes in emotional behavior, problem solving, and decision making. The brains of the mice that experienced adversity may be dysfunctional, in line with evidence of atrophy of frontal neurons after stress. Alternatively, humans and rodents may show positive adaptation to adversity. In one model currently growing in popularity, a brain developing under adversity may adopt a "live fast–die young" strategy favoring earlier maturation and short-term decision making. In this adaptive calibration model, genetically identical animals might express different sets of genes in their brains, and develop different neural circuits, in an attempt to prepare their brains for success in the kind of environments in which they found themselves. It is unclear how many possible trajectories might be available or when or how young brains are integrating environmental information. However, based on this adaptive calibration model, in lean times versus fat times you might expect a single species to "wire-up" different brains and display different behaviors without genetic change or genetic selection at the germline level.

Why should we care? This data might explain population-level economic behavior and offer a powerful counternarrative to seductive genetic explanations of success. Another piece of captivating news out recently was Nicholas Wade's review of Garett Jones's *Hive Mind* in the *Wall Street Journal*. Reading this review, I learned that national savings rates correlate with average IQ scores even if individual IQ scores do not; the book's subtitle is "How Your Nation's IQ Matters So Much More Than Your Own." In his review, Wade suggests that we need to look to "evolutionary forces" to explain IQ and its correlated behavioral differences. The well-controlled data from experiments in mice and rats suggest that we also look to early-life experience. Rodents are not known for their high IQ, but the bottom line is that what intelligence they have is sensitive to early-life experience even when we hold germline genetics constant. Putting these rodent and human findings together, one might hypothesize that humans exposed to instability or scarcity in early life are developing brains wired for shorter-term investment and less saving. Thus, rather than attributing the successes and failures of nations to slow-changing genetic inheritance, we might foster a brighter future by paying more attention to the quality of early-life experiences.

WE'RE STILL BEHIND

MARY CATHERINE BATESON

Professor emerita, George Mason University; visiting scholar, Sloan Center on Aging and Work, Boston College; author, *Composing a Further Life*

On October 4, 1957, the Soviet Union launched *Sputnik*, the planet's first artificial satellite. That was very big news, the beginning of an era of space exploration involving multiple launchings and satellites spending long periods in orbit, launched from many nations. In the weeks after *Sputnik*, however, another news story played out that led to a range of other actions based on the recognition that U.S. education was falling behind not only in science but in other fields of education as well, such as geography and foreign languages. This is still true. We are not behind at the cutting edge, but we are behind in general broad-based understanding of science, and this is not tolerable for a democracy in an increasingly technological world.

The most significant example is climate change. It turns out, for instance, that many basic terms are unintelligible to newspaper readers. Recently I encountered the statement that "a theory is just a guess—and that includes evolution," not to mention most of what has been reconstructed by cosmologists about the formation of the universe. When new data is published that involves a correction or expansion of earlier work, this is taken to indicate weakness rather than the great strength of scientific work as an open system, always subject to correction by new information. When the winter temperature dips below freezing, you hear, "This proves that the Earth is not warming." Most Americans are not clear on the difference between "weather" and "climate." The U.S. government supports the world's most advanced research on climate, but the funds to do so are held hostage by politicians convinced that climate change is a hoax. And we can add trickle-down economics and theories of racial and gender inferiority to the list of popular prejudices that many Americans believe are ratified by science.

Among the popular misconceptions of scientific concepts is a skewed concept of cybernetics as dealing only with computers. It is true that key concepts developed in the field of cybernetics resulted in computers as an important by-product, but the more significant achievement of cybernetics was a new way of thinking about causation, now more generally referred to as systems theory. Listen to the speeches of politicians proclaiming their intent to solve problems like terrorism. It's like asking for a single pill that will "cure" old age. If you don't like x, you look for an action that will eliminate it, without regard to the side effects (bombing ISIL increases hostility, for example) or the effects on the user (consider torture). Decisions made with overly simple models of cause and effect are both dangerous and unethical.

The news that has stayed news is that American teaching of science is still in trouble, and that errors of grave significance are made based on overly simple ideas of cause and effect, all too often exploited and amplified by politicians.

NEURAL HACKING, HANDPRINTS, AND THE EMPATHY DEFICIT

DANIEL GOLEMAN

Psychologist, science journalist; author, *A Force for Good: The Dalai Lama's Vision for Our World*

When I worked as a journalist at the science desk of the *New York Times*, our editors were constantly asking us to propose story ideas that were new, important, and compelling. The potential topics in science news are countless, from genetics to quantum physics. But if I were at the *Times* today, I'd pitch three science stories, all of which are currently under the collective radar, and each of which continues to unfold and will have mounting significance for our lives in years ahead.

For one: epigenetics. With the human genome mapped, the next step has been figuring out how it works, including what turns all those bits of genetic code on and off. Here everything from our metabolism to our diet to our environment and habits comes into play. A case in point is neuroplasticity. First considered seriously a decade or so ago, neuroplasticity—the brain's constant reshaping through repeated experiences—presents a potential for neural hacking apps. As neuroscientists like Judson Brewer at Yale and Richard Davidson at the University of Wisconsin-Madison have shown, we can choose which elements of brain function we want to strengthen through sustained mind training. Do you want to better regulate your emotions, enhance your concentration and memory, become more compassionate? Each of these goals means strengthening distinct neural circuitry through specific, bespoke mental exercise, which might one day become a new kind of daily fitness routine.

For another: industrial ecology as a technological fix. This new discipline integrates such fields as physics, biochemistry, and environmental science with industrial design and engineering to create a new method—life-cycle assessment (or LCA)—for measuring the ecological costs of our materialism. LCA gives a hard metric for how something as ubiquitous as a mobile phone affects the environment and public health at every stage in its life cycle. This methodology gives us a fine-grained lens on how human activities degrade the global systems that support life and points to specific changes that would bring the most benefit. Some companies are using LCA to change how their products are made, so that they will replenish rather than deplete. As work at the Harvard T. H. Chan School of Public Health illustrates, this means using LCA to shift away from the footprint metric (how much damage we do to the planet) to the handprint—measuring the good we do, or how much we reduce our footprint. A news peg: Companies are about to release the first major net-positive products, which, over their entire life cycle, replenish rather than deplete.

Finally: the inverse relationship between power and social awareness, which integrates psychology into political science and sociology. Ongoing research at the University of California at Berkeley by psychologist Dacher Keltner, and at other research centers around the world, shows that people who are higher in "social power"—through wealth, status, rank, or the like—pay less attention, in face-to-face encounters, to those who hold less power. Lessened attention means lessened empathy and understanding. Thus, those who wield power (such as wealthy politicians) have virtually no sense of how their decisions affect the powerless. Movements like Occupy, Black Lives Matter, and the failed Arab Spring can be read as attempts to overcome this divide. Such an empathy deficit will augment political tensions far into the future. Unless, perhaps, those in power follow Gandhi's dictate to consider how their decisions affect "the poorest of the poor."

SEND IN THE DRONES

DIANA REISS

Professor, Department of Psychology, Hunter College; author, *The Dolphin in the Mirror*

The increasing use of drone technology is revolutionizing wild-life science and changing the kinds of things we can observe. As a marine mammal scientist who studies cetaceans—dolphins and whales—I see how drones afford extended perception, far less intrusive means of observing and documenting animal behavior, and new approaches to protecting wildlife. Drones (referred to more formally as UAVs, for unmanned aerial vehicles) bring with them a new set of remote-sensing and data-collection capabilities.

The Holy Grail of observing animals in the wild is not being there, because your very presence is often a disturbing influence. Drones are a solution to this problem. Imagine the feeling of exhilaration and presence as your drone soars above a socializing pod of whales or dolphins, enabling you to spy on them from on high. We can now witness much of what was the secret life of these magnificent mammals. Myriad behaviors and nuances of interactions that could not be seen from a research boat—or would have been interrupted by its approach—are now observable.

Animal health assessments and animal rescues are being conducted by veterinarians and researchers with the aid of drones. For example, the Whalecopter, a small drone developed by research scientists at Woods Hole Oceanographic Institution in

Massachusetts, took high-resolution photographs of whales to document fat levels and skin lesions and then hovered in at closer range to collect samples of whale breath to study bacteria and fungi in their blow. NOAA scientists in Alaska are using drones to help them monitor beluga-whale strandings in Cook Inlet, providing critical information about the animals' condition, location, number, relative age, and whether they are submerged or partially stranded. The relayed images from drones are often clearer than those obtained by a traditional aerial surveys. Even if a drone cannot save an individual whale, getting more rapidly to a doomed whale enables scientists to conduct a necropsy on fresh tissue and determine the cause of death, which could bring about the survival of other whales.

Patrol drones are already being used to monitor and protect wildlife from poachers. One organization, Air Shepherd, has been deploying drones in Africa to locate poachers seeking elephant ivory and rhino horns. Programmed drones monitor high-traffic areas where the animals are known to congregate—areas known also to the poachers. They have been effective in locating poachers and informing the authorities of their whereabouts.

This is a new era of wildlife observation and monitoring. In my field, a future generation of cetacean-seeking drones may be around the corner—drones programmed to find cetacean-shaped forms and follow them. I can envision using a small fleet of "journalist drones" to monitor and provide realtime video feeds on the welfare of various species in our oceans, on our savannahs, and in our jungles. We might call it Whole World Watching (WWW) and create a global awareness, a more immediate connection between the world's human population and the other species sharing our planet.

THAT DRESS

SUSAN BLACKMORE

Psychologist; author, *Consciousness: An Introduction*

Could the color of a cheap dress create a meaningful scientific controversy? In 2015, a striped, body-hugging, £50 dress did just that. In February, Scottish mother Cecilia Bleasdale sent her family a poor-quality photo of a dress she bought for her daughter's wedding. Looking at the image, some people saw the stripes as blue and black, others as white and gold. Quickly posted online, "that dress" was soon mentioned nearly half a million times. This simple photo had everything a meme needs to thrive: It was easy to pass on, accessible to all, and sharply divided opinions. #thedress was, indeed, called the meme of the year and even a "viral singularity." Yet it did not die out as fast as it had risen; unlike most viral memes, this one prompted deeper and more interesting questions.

Scientists quickly picked up on the dispute and garnered some facts. Seen in daylight, the actual dress is indisputably blue and black. It is only in the slightly bleached-out photograph that white and gold is seen. In a study of 1,400 respondents who'd never seen the photo before, 57 percent saw blue and black, 30 percent saw white and gold, and about 10 percent saw blue and brown. Women and older people more often saw white and gold.

This difference is not like disputes over whether the wallpaper is green or blue. Nor is it like ambiguous figures, such as the famous Necker cube, which appears tilted toward or away from the viewer, or the duck/rabbit or wife/mother-in-law drawings. People typically see these bi-stable images either way and flip their perception between views, getting quicker with practice. Not so with "that dress." Only about 10 percent of people could switch colors. Most saw the colors, resolutely, one way only and remained convinced they were right. What was going on became a genuinely interesting question for the science of color vision.

Vision science has long shown that color is not the property of an object, even though we speak of it as though it were. Color in fact emerges from a combination of the wavelengths of light emitted or reflected from an object and the kind of visual system looking at it. A normal human visual system, with three cone types in the retina, concludes "yellow" when any one of an indefinite number of different wavelength combinations affects its color-opponent system in a certain way. Thus a species with more cone types, such as the mantis shrimp, which has about sixteen types, would see many different colors where humans would see only the same shade of yellow.

When people are red/green color-blind, with only two cone types instead of three, we may be tempted to think they fail to see something's "real" color. Yet there is no such thing. There are even rare people (mostly women) who have four cone types. Presumably they can see colors the rest of us cannot even imagine. This may help us accept the conclusion that the dress is not intrinsically one color or the other, but it still provides no clue as to why people see the dress so differently.

Could the background in the photo be relevant? In the 1970s, Edwin Land, inventor of the Polaroid camera, showed that the same colored square appears a different color depending on the squares surrounding it. This relates to an important problem that evolution has had to solve. If color information is to be useful, an object must look the same color on a bright sunny day as on an overcast one, yet the incident light is yellower at midday and bluer from a gloomy or evening sky. So our visual systems use a broad view of the scene to assess the incident light and then discount that when making color decisions, just like the automatic white balance (AWB) in modern cameras.

This, it turns out, may solve the Great Dress Puzzle. It seems that some people take the incident light as yellowish, discounting the yellow to see blue and black, while others assume a bluer incident light and see the dress as white and gold. Do the age and sex differences provide any clues as to why? Are genes, or people's lifetime experiences, relevant? The controversy is still stimulating more questions.

Was it a step too far when some articles suggested that #thedress could prompt a "worldwide existential crisis" over the nature of reality? Not at all, for color perception really is strange. When philosophers ponder the mysteries of consciousness, they may refer to qualia—private, subjective qualities of experience. An enduring example is "the redness of red," because the experience of seeing color provokes all those questions that make the study of consciousness so difficult. Is someone else's red like mine? How could I find out? And why, when all this extraordinary neural machinery is doing its job, is there subjective experience at all? I would guess that "that dress" has yet more fun to provide.

ANTHROPIC CAPITALISM AND THE NEW GIMMICK ECONOMY

ERIC R. WEINSTEIN

Mathematician and economist; managing director, Thiel Capital

Consider a thought experiment: If market capitalism was the brief product of happy coincidences confined in space and time to the developed world of the 19th and 20th centuries (but no longer held under 21st-century technology), what would our world look like if there were no system to take its place? I have been reluctantly forced to the conclusion that if technology had killed capitalism, economic news would be indistinguishable from today's feed.

Economic theory, like the physics on which it's based, is in essence an extended exercise in perturbation theory. Solvable and simplified frictionless markets are populated by rational agents, which are then all subjected to perturbations in an effort to recover economic realism. Thus, while economists do not, as outsiders contend, believe idealized models to be exactly accurate, it's fair to say that they assume deviations from the ideal are manageably small. Let's list a few such heuristics that may have recently been approximately accurate but aren't enforced by any known law:

- Wages set to the marginal product of labor are roughly equal to the need to consume at a societally acceptable level.

- Price is nearly equal to value, except in rare edge cases of market failure.
- Prices and outputs fluctuate coherently so that it's meaningful to talk of scalar rates of inflation and growth (rather than varying field concepts like temperature or humidity).
- Growth can be both high and stable, with minimal interference by central banks.

The anthropic viewpoint (more common in physics than economics) on such heuristics would lead us to ask, "Is society now focused on market capitalism because it is a fundamental theory, or because we have just lived through the era in which it was possible due to remarkable coincidences?"

To begin to see the problem, recall that in previous eras innovations created high-value occupations by automating or obviating those of lower value. This led to a heuristic that those who fear innovation do so because of a failure to appreciate newer opportunities. Software, however, is different in this regard, and the basic issue is familiar to any programmer who has used a debugger. Computer programs, like life itself, can be decomposed into two types of components: (1) loops, which repeat with small variations, and (2) Rube Goldberg–like processes, which happen once.

If you randomly pause a computer program, you'll almost certainly land in the former, because the repetitive elements are what gives software its power, by dominating the running time of almost all programs. Unfortunately, our skilled labor and professions currently look more like the former than the latter, which puts our educational system in the crosshairs of what software does brilliantly.

In short, what today's flexible software is threatening is to "free" us from the drudgery of *all* repetitive tasks rather than those of lowest value, pushing us away from expertise, which we know how to impart, toward ingenious Rube Goldberg–like opportunities unsupported by any proven educational model. This shift in emphasis from jobs to opportunities is great news for a tiny number of today's creatives but troubling for a majority who depend on stable and cyclical work to feed families. The opportunities of the future should be many and lavishly rewarded, but it's unlikely they will ever return in the form of stable jobs.

A next problem is that software replaces physical objects by small computer files. Such files have the twin attributes of what economists call public goods: (1) The good must be inexhaustible (my use doesn't preclude your use or reuse), and (2) the good must be non-excludable (the existence of the good means that everyone can benefit from it even if they don't pay for it).

Even die-hard proponents of market capitalism will cede that this public sector represents "market failure" where price and value become disconnected. Why should one elect to pay for an army when he will equally benefit from free-riding on the payments of others? Thus in a traditional market economy, payment must be secured by threat of force, in the form of compulsory taxes.

As long as public goods make up a minority of a market economy, taxes on nonpublic goods can be used to pay for the exception where price and value gap. But in the modern era, things made of atoms (e.g., vinyl albums) are being replaced by things made of bits (e.g., MP3 files). While 3D printing is still immature, it vividly showcases how the plans for an object will allow us to disintermediate its manufacturer. Hence, the previous edge case of market failure should be expected to claim an increasingly dominant share of the pie.

Assuming that a suite of such anthropic arguments can be made rigorous, what will this mean? In the first place, we should expect that because there is as yet no known alternative to market capitalism, central banks and government agencies publishing official statistics will be under increased pressure to keep up the illusion that market capitalism is recovering, by manipulating whatever dials can be turned by law or fiat, giving birth to an interim "gimmick economy."

If you look at your news feed, you'll notice that the economic news already no longer makes much sense in traditional terms. We have strong growth without wage increases. Using Orwellian terms like "Quantitative Easing" or "Troubled Asset Relief," central banks print money and transfer wealth to avoid the market's verdict. Advertising and privacy transfer (rather than user fees) have become the business model of last resort for the Internet corporate giants. Highly trained doctors squeezed between expert systems and no-frills providers are morphing from secure professionals into service-sector workers.

Capitalism and communism, which briefly resembled victor and vanquished, increasingly look more like Thelma and Louise—a tragic couple sent over the edge by forces beyond their control. What comes next is anyone's guess, and the world hangs in the balance.

THE ORIGIN OF EUROPEANS

GREGORY COCHRAN

Physicist; adjunct professor of anthropology, University of Utah;
co-author (with Henry Harpending), *The 10,000 Year Explosion:
How Civilization Accelerated Human Evolution*

Europeans, as it turns out, are the fusion of three peoples—
blue-eyed, dark-skinned Mesolithic hunter-gatherers, Anatolian
farmers, and Indo-Europeans from Southern Russia. The first
farmers largely replaced the hunters (with some admixture) all
over Europe, so that 6,000 years ago populations from Greece
to Ireland were genetically similar to modern Sardinians—dark-
haired, dark-eyed, light-skinned. They probably all spoke related
languages, of which Basque is the only survivor.

About 5,000 years ago, Indo-Europeans arrived out of the
East, raising hell and cattle. At least some of them were probably blond or red-headed. In northern Europe they replaced
those first farmers, root and branch. Germany had been dotted
with small villages before their arrival—immediately afterward, no buildings. Mitochondrial variants carried by one in
four of those first farmers are carried by 1 in 400 Europeans
today, and the then-dominant Y chromosomes are now found at
the few-percent level on islands and mountain valleys—refugia.
It couldn't have been pretty. In Southern Europe, the Indo-
Europeans conquered and imposed their languages, but without
exterminating the locals—even today Southern Europeans are
mostly descended from those early farmers.

In other words, the linguists were correct. For a while, the archeologists were, too: V. Gordon Childe laid out the right general picture (*The Aryans: A Study of Indo-European Origins*) back in 1926. But then progress happened: Vast improvement in archeological techniques, such as C-14 dating, were accompanied by vast decreases in common sense. Movements of whole peoples—invasions and *Völkerwanderungs*—became "problematic," unfashionable: They bothered archeologists and *therefore must not have happened*. Sound familiar?

The picture is clear now, due to investigations of ancient DNA. We can see whether populations are related or not, whether they fused, or whether one replaced another and to what extent. We even know that one group of ancient Siberians contributed to both Indo-Europeans and Amerindians.

We also know that modern social scientists are getting better and better at coming to false conclusions. You could blame the inherent difficulty of a historical science like archeology, where experiments are impossible. You might blame well-funded STEM disciplines for drawing away many of the sharper students. You could blame ideological uniformity—but you would be mistaken. Time travelers bringing back digitally authenticated full-color 3D movies of prehistory wouldn't fix this problem.

Their minds ain't right.

THE PLATINUM RULE: DENSE, HEAVY, BUT WORTH IT

HAZEL ROSE MARKUS

Davis-Brack Professor in the Behavioral Sciences, Stanford University; co-editor, *Doing Race: 21 Essays for the 21st Century*

The variously attributed Platinum Rule holds that we should do unto others as *they* would have us do unto *them*. The most important news is that there is growing evidence that every endeavor involving social connection—friendship and marriage, education, healthcare, organizational leadership, interracial relationships, and international aid, to name a few—is more effective to the extent that it adheres to this behavioral guide. The reason that the beneficial consequences of holding to the rule will remain important news is that the Platinum Rule is not simple and hewing to it is tough, especially in an individualist culture that fosters the wisdom of one's own take on reality. Following the dictates of the Platinum Rule is so tough, in fact, that we routinely ignore it and then find it surprising and newsworthy when a new study discovers its truth all over again.

The challenge of holding to the Platinum Rule begins with the realization that it is not the Golden Rule—do unto others as *you* would have them do unto *you*. The Golden Rule is also a good behavioral guide and one that shows up across the religious traditions (e.g., in Judaism—what is hateful to you, do not do to your neighbor; in Confucianism—do not do to others what you do not want done to yourself). Yet built into the foundation of the Golden Rule is the assumption that what is good, desirable, just, respectful, and helpful for *me* will also be good, desirable, just, respectful, and helpful for *you* (or should be, and if it isn't right now, trust me, it will be eventually).

Even with good friends or partners, this is often not the case. For example, from your perspective, you may be certain you're giving me support and fixing my problem. Yet what I would prefer and find supportive and *have you do unto me* is for you to listen to me and my analysis of the problem. In the many cases in which we strive to connect with people across social class, sexual orientation, race, ethnicity, religion, and geography, there is almost certainly some disconnect between how you think you should treat people and how they would like to be treated. Doing unto others as they would have you do unto them requires knowledge of others, their history and circumstances, what matters to them. It means appreciating and acknowledging the value of difference and accommodating one's actions accordingly.

At the base of the successful application of the Platinum Rule is the realization that one's own way may not be the only or the best way. Yes, not all ways are good; some are uninformed, corrupt, and evil. Yet the findings from cultural science are increasingly robust: There is more than one good or right or moral way to raise a child, educate a student, cope with adversity, motivate a workforce, develop an economy, build a democracy, be healthy, and experience well-being.

What is good for me and what I assume will be good for you too is likely grounded in what cultural psychologists call a WEIRD (Western, Educated, Industrialized, Rich and Democratic) perspective. For the 75 percent of the world's population who cannot be so classified (a majority that includes many people in North America without a college degree or with non-

Western heritage), who I am, what matters to me, what I hope to be, and what I would most like done unto me may not match what seems so obviously and naturally good and appropriate from the WEIRD perspective.

Besides knowledge of the other and appreciation of the difference, the Platinum Rule requires something harder—holding one's own perspective at bay while thinking and feeling one's way into the position of the other and then creating space for this perspective. Such effort requires a confluence of cognitive, affective, and motivational forces. Some researchers call this psychological work perspective-taking; some call it empathy, some compassion, still others social or emotional intelligence. Whatever the label, the results are worth the effort.

When colleges ask students from working-class or under-represented-minority backgrounds to write about what matters to them or to give voice to their worries about not fitting in in college, they are happier, healthier, and outperform students not given these opportunities. Managers who encourage employees to reflect on the purpose and meaning of their work have more effective teams than those who don't. The odds of persuading another in an argument are greater if you acknowledge the opponent's moral position before asserting your own. Research from across the social sciences supports the idea that just recognizing the views, values, needs, wants, hopes, or fears of others can produce better teaching, medicine, policing, team leadership, and conflict resolution. Taking others' views into account may change the world.

Perhaps even more newsworthy than the successes of understanding what matters to others are the many failures to apply the Platinum Rule. Government and private donors distributed billions following the 2010 earthquake in Haiti, much of it spent doing what the donors believed they would do if they were in the place of those devastated by the disaster. One notable project was a campaign to underscore the good health consequences of handwashing to people without soap or running water. Many humanitarians now argue that relief efforts would be less costly and more effective if instead of giving people what donors think they need—water, food, first-aid kits, blankets, training—they delivered what the recipients themselves say they need. In most cases, this is money.

Whether independent North Americans (who, according to some surveys, are becoming more self-focused by the year) can learn the value of the Platinum Rule is an open question. At this point, the science suggests that it would be moral, efficient, and wise.

ADJUSTING TO FEATHERED DINOSAURS

JOHN McWHORTER

Professor of linguistics, Columbia University; cultural commentator; author, *The Language Hoax*

The discovery that dinosaurs of the Velociraptor type had ample feathers and looked more like ostriches than the slick beasts in the *Jurassic Park* movies was my favorite scientific finding of 2015.

The specific discovery was the genus Zhenyuanlong, but it tells a larger story. Feathered dinosaurs have been coming out of the ground in China almost faster than anyone can name them, since the 1990s. However, it has taken us a while to recognize that the feathers on these dinosaurs mean that equivalent dinosaurs in other parts of the world had feathers as well. The conditions in China simply happen to have been uniquely suited to preserving the feathers' impressions. That paleontologists now know that even Velociraptor had feathers is a kind of unofficial turning point. No longer can we think of feathered dinosaurs as a queer development of creatures in East Asia. We can be sure that classic dinosaurs of that body type, traditionally illustrated with scaly lizard-type skins—dino fans of my vintage will recall Coelophysis, Ornitholestes, etc.—had feathers. Other evidence of this kind came in 2015 too, including the discovery of strikingly extensive evidence of feathers on dinosaurs long known as the "ostrich-like" sort—Ornithomimus. Little did we know how close that resemblance was.

"Who really cares whether Velociraptor had feathers?" one might ask. But one of the key joys of science is discovering the unexpected. One becomes a dino fan as a kid, from the baseball-card-collecting impulse, savoring one's mental list or cupboard of names and types. However, since the seventies it has become clear, first, that birds are the dinosaurs that survived, but then, more dramatically, that a great many dinosaurs had feathers just like birds. This shouldn't be surprising, but it is. Dinosaurs have gone from hobby to mental workout.

Second, the feathered Velociraptor coaxes us to tease apart the viscerally attractive from the empirically sound. The truth is that the sleek versions of bipedal dinosaurs look "cool"—streamlined, shiny, reptilian in a good way. Nothing has made this clearer than the Velociraptors brought to life in the *Jurassic Park* films. For these creatures to instead look more like ostriches, sloshing their feathers around and looking vaguely uncomfortable, doesn't quite square with how we're used to seeing dinosaurs. Yet that's the way it was. The sleek Velociraptor or any number of other dinosaurs of that general build are now "old school," just like Brontosauruses lolling around in swamps (they didn't) and Tyrannosauruses dragging their tails on the ground.

Finally, Velociraptor as ostrich neatly reinforces for us the fact that evolution works in small steps, each of which is functional and advantageous at the time, but for reasons that can seem quite disconnected from the purpose of the current manifestation of the trait. In life in general, this is a valuable lesson.

In a language like Spanish that marks articles, adjectives, and nouns with an arbitrary gender, the gender-marking helps clarify how the words relate to one another. But such marking originates from a division of nouns into such classifications as "masculine" and "feminine" (or animal, long, flat, etc., depending on the language) whose literal meanings fade over time, leaving faceless markers.

In the same way, an ostrich-sized dinosaur with feathers clearly wasn't capable of soaring like an albatross, which is one of many pieces of evidence that feathers emerged as insulation and/or sexual display and only later evolved to allow flight.

Not that dinosaurs existed to conform to the aesthetics, tastes, and nostalgic impulses of observers millions of years after their demise, but feathered dinosaurs are tough to adjust to. Yet the adjustment is worth it: It makes dinosaurs more genuinely educational in many ways.

PEOPLE ARE ANIMALS

LAURA BETZIG

Anthropologist; historian; author, *Despotism and Differential Reproduction: A Darwinian View of History*

As an aphorism, that isn't news at all. 2,300-plus years ago, Aristotle wrote in his *Politics*: "It is evident that the state is a creation of nature, and that man is by nature a political animal." He thought we were even more political than bees.

But that apothegm became science after Darwin traced *Homo sapiens'* descent from apes, and generations of Darwin's scientific descendants—especially the last generation—followed up with field studies of hundreds of other social, or "political," animals. They showed that whenever animals get together, they play by the same rules. When groups form in vaguely delineated habitats, where the costs of emigrating are low, they tend to be quasi-social. Most animals help raise the young and most animals reproduce. But when groups form in sharply delineated habitats, where the costs of emigrating are high—say, in the tree hollows where thousands of honeybees raise larvae produced by their queen—they're often eu-(or "truly") social. Some animals are breeders. Other animals work.

As long as they kept moving, most *Homo sapiens* probably raised their own children. For roughly 100,000 years, they ran around the sub-Sahara; then, around 100,000 years ago, they left Africa. After around 10,000 years ago, they settled down in the Fertile Crescent and their societies started to look like those of the bees.

Their hagiographers had medieval missionaries soak up chastity like honey, and some of the abbots who worked under Charlemagne were referred to as *apis*. Saint Ambrose, who venerated virginity, was discovered in his Trier cradle with a swarm of bees in his mouth and ended up as a honey-tongued bishop; Saint Bernard, who founded a monastery in Burgundy and venerated the Virgin Mary, was remembered as Doctor Mellifluus (the Honey-Sweet Doctor). But most helpers in the Middle Ages retained their genitals.

Workers in the first civilizations did not. There were apiaries in ancient Israel, the land of milk and honey, where David made his son Solomon a king in front of an assembly of eunuchs. And there were apiaries in Egypt, where pharaohs put bees on their cartouches and may have collected civil servants who lacked generative powers. There were more beekeepers in Mantua, where Publius Vergilius Maro grew up. For the benefit of the first Roman emperor, Augustus, he remembered how honeybees fought and foraged for their monarch, made honey, and nursed their larvae. In the end, unmarried soldiers, unmarried slaves, and the unmanned attendants of the sacred bedchamber effectively ran that empire. And the emperor bred.

The take-home message from all of which is simple. It's good to be mobile. Societies, human or otherwise, are politically and reproductively more equal when emigration is an option. They're less equal when that option is closed.

THE LONGEVITY OF NEWS

DIANA DEUTSCH

Professor of psychology, UC San Diego; author, *The Psychology of Music*

A remarkable thing about any piece of news, scientific or otherwise, is that it's hard to gauge its longevity. A prime example of "important" scientific news that turned out to be mistaken is the "discovery" of N-rays by the physicist Prosper-René Blondlot in 1903. This was hailed as a major breakthrough and led rapidly to the publication of dozens of papers claiming to confirm Blondlot's findings. Yet N-rays were soon discredited and are now referred to primarily as an example of a phenomenon in perceptual psychology: We perceive what we expect to perceive.

On the other hand, scientists often underrate the practical importance of their discoveries, so that news about them does not begin to do justice to their implications. When Edison patented the phonograph in 1878, he believed it would be used primarily for speech, such as for dictation without the aid of a stenographer, for books that would speak to blind people, for telephone conversations that could be recorded, and so on. Only later did entrepreneurs realize the enormous value of recorded music. Once they did, the music industry developed rapidly.

The laser is another example of the underrating of the practical implications of a scientific discovery. When Schawlow and Townes published their seminal paper describing the principle of the laser in *Physical Review* in 1958, it produced considerable excitement in the scientific community and eventually won them Nobel Prizes. However, neither these authors nor others in their group predicted the enormous and diverse practical implications of their discovery. Lasers, apart from their many uses in science, have enabled the development of fast computers, target designation in warfare, communication over very long distances, space exploration and travel, surgery to remove brain tumors, and numerous everyday uses—bar-code scanners in supermarkets, for example. Arthur Schawlow frequently expressed strong doubts about the laser's practicality and often quipped that it would be useful only to burglars for safecracking. Yet advances in laser technology continue to make news to this day.

WEATHER PREDICTION HAS QUIETLY GOTTEN BETTER

SAMUEL ARBESMAN

Complexity scientist; scientist in residence, Lux Capital;
author, *Overcomplicated: Technology at the Limits of Comprehension*

Surveying the landscape of scientific and technological change, one sees a number of small advances that together have unobtrusively yielded something startling. Through a combination of computer-hardware advances (Moore's Law marching on), ever more sophisticated algorithms for solving certain mathematical challenges, and larger amounts of data, we have got something new—really good weather prediction. According to a recent paper in *Nature*,

> Advances in numerical weather prediction represent a quiet revolution because they have resulted from a steady accumulation of scientific knowledge and technological advances over many years that, with only a few exceptions, have not been associated with the aura of fundamental physics breakthroughs.*

Despite their profound unsexiness, these predictive systems have yielded enormous progress. Our skill at forecasting the weather several days ahead has increased in accuracy by about an additional day per decade over the past few decades and

---

* http://www.nature.com/nature/journal/v525/n7567/abs/nature14956.html

changed how we think about the weather's enormously complex system.

This is important for a number of reasons. Understanding weather is vital for a huge number of human activities, from transportation to improving agricultural output to managing disasters. But there's a potentially bigger reason. While I'm hesitant to extrapolate from the weather system to other complex systems—including many that are perhaps much more complex, such as living organisms or entire ecosystems—this development should give us some hope. That weather prediction has improved via technological advancement and scientific and modeling innovations means that other problems we might deem unsolvable needn't be. This news of a "quiet revolution" in weather prediction might be a touchstone for how to think about predicting and understanding complex systems. Never say never when it comes to intractability.

THE WORD: FIRST AS ART, THEN AS SCIENCE

BRIAN CHRISTIAN

Philosopher; computer scientist; poet; co-author (with Tom Griffiths),
Algoriths to Live By

The phrase "news that stays news" was originally how Ezra
Pound, in 1934, defined literature—and so it's interesting to
contemplate what, in the sciences, might meet that standard. The
answer might be the emerging science of literature itself.

Thinking about the means by which language works on
the mind, Pound described a three-part taxonomy. First is
phanopoeia—think "phantoms," the images that a word or phrase
conjures in the reader's mind. Pound's "petals on a wet black
bough" is a perfect illustration. Phanopoeia, he says, is the poetic
capacity most likely to survive translation. Second is *melopoeia*—
think "melody," the music words make. This encompasses rhyme
and meter, alliteration and assonance, the things we take to be
the classic backbones of poetic form. Though fiendishly difficult
to translate faithfully, he notes, it doesn't necessarily need to be,
as this is the poetic capacity most likely to be appreciated even
in a language you don't know.

Third and most enigmatic is a quality Pound called *logopoeia*
and described as "akin to nothing but language," "a dance of
the intelligence among words." This has proved the most elusive to characterize, but Pound later noted that he meant something like verbal register, the unique patterns of usage unique to
each word. Take a pair of words like "doo" and "stool." They

can both denote the same thing; their sonic effects are about as near as any pair of words can be. And yet their difference in register—one juvenile, the other clinical—is so strong that the words can't even be considered synonyms, as it's almost impossible to imagine a context in which one could be substituted for the other.

Logopoeia proves to be one of the most dazzling of poetic effects—see, for instance, the contemporary poet Ben Lerner, who writes lines like "a beauty incommensurate with syntax had whupped my cracker ass"—but also the most fragile. It's almost impossible to translate faithfully, because every language divides its register space differently. See for instance the French film *The Class (Entre les Murs)*, in which a teacher tells a pair of students they were behaving with *"une attitude de pétasses."* The English version subtitled the line "acting like skanks" and prompted a minor furore over whether that particular word was stern enough to serve as an admonishment that would get through to an unruly student, yet inoffensive enough for a teacher to say without expecting to jeopardize his job, yet offensive enough to do exactly that. What's more, an entire scene pivots on the fact that for the students the word strongly implies "prostitute" whereas for the teacher it has no such pointed connotation. What word in English meets all those criteria? Maybe there *is* no such word in English.

Logopoeia, in fact, is so fragile that it doesn't even survive in its *own* language for long. The *New York Times* included the word "scumbag" in a crossword puzzle in 2006, a word almost charmingly inoffensive to their editorial staff and the majority of the public but jaw-droppingly inappropriate to readers old enough to remember the word when it couldn't be spoken in polite company, as it explicitly summoned the image of a used condom. Changes like this are everywhere in a living language. In 1990, it would have been unthinkable for my parents to say "Yo," for instance. In 2000, when they said it, it was painful and tone-deaf, a sad attempt to sound like a younger and/or cooler generation. By 2010, it was just about as normal as "Hey." How could a reader (let alone a translator) some centuries hence possibly be expected to know the logopoetic freight of every single word at the time of the piece's writing?

For the first time in human history, we have the tools to answer this question. A century after logopoeia entered the humanities, it is becoming a science. Computational linguists now have access to corpora large enough, and computational means sufficient, to see these forces in action: to observe words as they emerge, mutate, and evolve; to quantify logopoeia, the subtlest and most ephemeral of linguistic effects.

This has changed our sense of what a word is. The question is far from academic. When the FCC moved to release a set of documents from a settlement with AT&T to the public in the mid-2000s, AT&T argued in court that this constituted "an unwarranted invasion of personal privacy," citing that it was a "legal person" in the eyes of the law. The Third Circuit, in 2009, agreed, and the FCC appealed. The case went to the Supreme Court to decide, in effect, whether "person" and "personal" are two forms of the same word or two independent terms that happen to share a lot of their orthography (and at least some of their sense).

The Court traditionally has turned to the Oxford English Dictionary in situations like this. In this instance, though, they turned instead to computational linguists, who performed an analysis across an enormous corpus of real-world usage to investigate whether the words are used in the same contexts, in the vicinity of the same words. The analysis determined that they are not. The words were shown to be divergent enough to constitute two independent terms; thus not every "person" is necessarily entitled to "personal privacy." The documents were released. "We trust," wrote Chief Justice John Roberts in the majority decision, "that AT&T will not take it personally."

The rapidly maturing science of computational linguistics, possible only in a Big Data era, has finally given scholars of the word what the telescope is to astronomy or the microscope to biology. That's big news.

And because words, more unstable than stars and squirrelier than paramecia, refuse to sit still, changing context subtly with every utterance, it's news that will stay so. Pound would, I think, agree.

THE CONVERGENCE OF IMAGES AND TECHNOLOGY

VICTORIA WYATT

Associate professor of the indigenous arts of North America,
University of Victoria, Canada

The news is in pictures, literally and figuratively. Visual images have exploded through our world, challenging the primacy of written text. A photograph bridges the diversity of cultures and languages. Tens of thousands of independent agents send it racing through overlapping networks. Public responses surge globally with exponential speed. Political leaders act, or fail to.

Never before have visual images so dynamically pervaded our daily lives. Never before have they been so influentially generated by amateurs as well as editors and advertisers. Digitization brings the creation of images within everyone's purview. The Internet gives us the means to communicate visually and the imperative to do so. Images now form a necessary component of even heavily text-based Web sites. Social media coalesce around visual imagery. Written text works brilliantly in many ways, but it has never worked in quite this way.

The convergence of technology and the visual does not announce itself with the éclat of a seminal scientific breakthrough. It claims no headlines. Our culture associates images with infancy. Pictures appear in childhood storybooks, disappearing as we progress to sophisticated novels. Our new emphasis on the image has much to surmount. In the future, some critics will condemn it as the tipping point in the death of literacy. They will be wrong. It *is* a tipping point and a stealthy one, but for a different reason. It lays the foundation for the paradigm shift essential to our survival.

Reading is a linear experience. Alphanumeric text unfolds inexorably in unidirectional, chronological sequence. It calls on us to focus narrowly on symbols in lines isolated from context. To read, we retreat from our hugely complex visual environment.

Granted, the content of written text can refer to complexities. It often does. Poetic prose can use rhymes and resonances to signal relationships and make meanings potent. Always, though, alphanumeric text comprises discrete segments, not holistic representations. We read words, sentences, paragraphs consecutively. We must gather them together ourselves to construct and consider the relationships therein. A visual image embodies the whole at a glance. All the intangible connections, all the invisible yet pregnant relationships between the component parts, present themselves in concert. It's up to us to perceive these intertwined threads and make meaning of them. Sometimes we do; sometimes we don't. Regardless, in the image they simultaneously exist.

This is how we live. We do not experience our world as a series of discrete visible components. Such distortion of reality would have compromised evolutionary success. We intuit the network of invisible relationships underlying the concrete entities we see, and create holistic meaning accordingly. Visual images prompt us to do the same.

Innovations in data visualization underscore the value of visual imagery in representing intangibles. Again, technology makes it possible. Computers find nonlinear patterns in space and time embedded in huge data sets. Programs such as spatial mapping make these complex connections vivid. Scientists have long used visualizations to portray natural systems. Increasingly, social and cultural researchers choose similar software to embody subjective human experience. Interactive maps show dynamic networks in process, not frozen instants of artificial stasis. As technology opens new avenues for exploration of relationships, disciplines across academia embrace fresh questions in emerging forms. To focus on intangibles, these questions demand the power of imagery.

A tsunami of visual images washes over our world. But "tsunami," though a visual metaphor, is a poor one, implying danger; rather, the new immersion in visual images counters a perilously segmented perspective. Written text is important in recent human history and will continue to be so, for obvious reasons. Authentically representing reality is not one of those reasons. Visual images gain such popularity and such currency today because they achieve what written text cannot: They show us the intangibles defining our world.

One might think the elevation of the image would prompt us all the more to focus solely on what we see; in fact, immersion in visual imagery mirrors how we experience reality—constantly constructing meaning from invisible relationships in our visual field. The famous metaphor about perception, "You can't see the forest for the trees," hints at this process but misses the greater paradigm shift. The forest remains a visible entity. We need to discern the invisible, intangible ecosystem that underlies our forest and drives all that happens there. Visual images remind us to look. They help us focus on what we *cannot* see.

Our future depends on how well we do that. Today's marriage of technology and the visual gives us the means. The Internet gives us popular demand. It mirrors the complexities of holistic visual experience comprising intangible connections. Even in digital text, highlighted hyperlinks bombard us with visual reminders that relationships exist. We explore Web connections in orders and directions of our own choosing. We receive information immediately thinking of whether and with whom to share it. A generation has grown up expecting assertive interaction with nonlinear formats. Technology paired with imagery frees us from the artificial isolation of linear reading. We will never return to that solitary confinement.

The news *is* in pictures. We do stand at a tipping point, created by the convergence of images and technology. In the future, this moment may be decried as the death knell for literacy, just another item in a long list of societal failings. Or it may be extolled as the popular vanguard of a paradigm that makes global problem-solving possible. What it will mean in twenty-five years depends on what we all make it mean now.

THE MINDFUL MEETING OF MINDS

CHRISTINE FINN

Archeologist; journalist; author, *Past Poetic: Archaeology and the Poetry of W. B. Yeats and Seamus Heaney*

Sometime around the end of the last century, a TV journalist I knew reported a news story on a school that was teaching meditation to its pupils. She was personally skeptical. The resulting piece was controversial. I gather the class was discontinued.

Fast forward to last year, when I came across a cover story in a national magazine about meditation which used an illustration of a blonde white woman in peaceful posture. The controversy this time was not about meditation per se but that the article was illustrated with an image of a cliché too far, it seemed; the audience was broader than that. Not the subject matter, then, but what was seen by some as a narrow portrayal of it.

Full disclosure: I am also a blonde white woman and I have practiced meditation for twenty years. For much of that time I hesitated to admit it. I came to it as a postgrad working on an unfunded interdisciplinary thesis. Fizzy with discoveries in the fuzzy zone, I needed to corral my brain if I was to defend my argument as both art and science. And somewhere along the way, science got more interested in meditation. So now I can openly discuss having had my brainwaves sampled and what the results looked like on a graph.

But my point here is not to make an argument for meditation. What I find interesting and newsworthy is the existence of a broadening dialogue between what was until recently a fringe subject and the rigorous realm of verification and repeatable experiment. C. P. Snow's 1959 argument about the gulf between science and the humanities hits home in this *Edge* context. The current blurring of lines is encouraged by online media, even as meditation is being investigated as a salve to the digital age.

In my example, a story about meditation reports the result of experiment, cites academic papers, draws conclusions, and suggests causes; the audience reports effects and experiential data from another form of experimentation—practice. The flow is as two-way as the attentive breathing at the center of meditation. And it has the potential to enlighten both scientist and practitioner. They can, of course, be one and the same.

That's what is newsworthy: the counterculture of science in the many stories (catch line, "Mindfulness") now streaming through the media at a confident pace. Those stories gathered in parks, prisons, offices, hospitals, nursing homes, hospices—and schools. Moving betwixt and between, and toward an interesting new stillness.

CARPE DIEM

ERNST PÖPPEL

Neuroscientist; cofounder and chairman, Human Science Center,
Ludwig-Maximilians-Universität, Munich; author, *Mindworks*

Some 2,000 years ago, probably 23 B.C.E., the Roman poet
Horace published some poems, and they lasted forever, as he
predicted himself: "I have built a monument more lasting than
bronze" (*Exegi monumentum aere perennius*). Although his words
lacked modesty, he was right. The most famous ode (number
11 of the first book) is also one of the shortest, with only eight
lines. Everyone knows the phrase "Enjoy the day" (*carpe diem*);
the Latin implies more than just having fun: to actively grasp the
opportunities of the day, to "seize the present."

This ode of Horace starts with the energetic advice not to
ask questions that cannot be answered. It's an eternal reminder,
and not only for scientists; it is very old news that has to be repeated regularly. Science's purpose is to discover right and good
questions—indeed, often unasked questions—before trying
to supply an answer. But what are criteria for right questions?
How do we know that a question can be answered and does not
belong to the realm of irrationality? How can a mathematician
trust that a proof will be possible before spending years to find
it? Apparently, the power of implicit knowledge, or intuition, is
much stronger than we are inclined to believe. Albert Einstein
is said to have remarked: "The intuitive mind is a sacred gift and
the rational mind is a faithful servant. We have created a society
that honors the servant and has forgotten the gift."

In poetry we often find representations of such implicit knowledge and intuition with high scientific value, opening new windows of potential discoveries. If read (even better, spoken) with an open mind, poetry can serve as a bridge, an effortless link, between the different cultures. Thus, poetry does not belong to the humanities only (if at all); poems in all languages express anthropological universals and cultural specifics in a unique way and provide insights into human nature, the mode of thinking and experiencing, often shadowed by a castrated scientific language.

After his warning with respect to questions that cannot be answered, Horace suggests that one should simply accept reality (*ut melius quidquid erit pati*), and he gives the frustrating but good advice to bring our great hopes into a smaller space (*spatio brevi spem longam reseces*). This is hard to take; scientists always want to go beyond the limits of our mental power. But it is good to be reminded that our evolutionary heritage has dictated limits of reasoning and insights which must be accepted and should instill modesty.

Other cultures, too, have pointed out such limits. More than 2,000 years ago, Lao Tzu, in the *Tao Te Ching*, says: "To know about not knowing is the highest" (in pinyin with indication of the tones: zhi-1 bu-4 zhi-1, shang-4). To accept such an attitude is not easy, and it may be impossible to suppress the search for causality, as expressed by the French poet Paul Verlaine: "It is the greatest pain not to know why" (*C'est bien la pire peine de ne savoir pourquoi*).

Apparently, poets (of course not all of them) have some knowledge about our mental machinery which can guide scientific endeavors. But there is also a problem, which is language itself: Can poetry be translated? Can even scientific language be translated veridically? Of course not. Take the English translation, "Seize the present," of *Carpe diem*. Does the English "present" cover equivalent connotations in Chinese, German, or any other language? The English "present" evokes different associations compared to the German *Gegenwart* or the Chinese *xianzai*. "Present" is associated with sensory representations, whereas *Gegenwart* has a more active flavor; the component *warten* refers either to "to take care of something" or "to wait for something," and it is thus also past- and future-oriented. The Chinese *xianzai* is associated with the experience of existence in which something is accessible by its perceptual identity; it implies a spatial reference, indicating the *here* as the locus of experience, and it is also action-oriented. Although the different semantic connections are usually not thought of explicitly, they still may create a bias within an implicit frame of reference.

What follows? We must realize that the language one uses, including that of scientific discourse, is not neutral with respect to what one wants to express. But this is not a limitation; if one knows several languages—and a scientist knows several languages anyway—it is a rich source of creativity. Some sentences, however, do not suffer from translations; they are easily understood and they last forever. When Horace says that while we talk, the envious time is running away (*dum loquimur, fugerit invida aetas*), one is reminded of scientific (and political) discussions full of words with not too much content—not exactly new news.

LINKING THE LEVELS OF HUMAN VARIATION

ELIZABETH WRIGLEY-FIELD

Assistant professor, Department of Sociology, University of Minnesota-Twin Cities

We are rewriting the story of human populations with data that depict individuals simultaneously from above and below: at scales geographic and genetic, from social networks to microbial networks. What is new is not the aspiration to integrate each level of human experience but the data that make it possible.

When we have only one kind of data, we can find only one kind of answer. But in the social sciences, explanations are like ecosystems. The presence of people—and their leftovers— enables mice to live in a house; the presence of a cat constrains them. Just as a species' niche expands or contracts with the presence of other species, explanatory factors, too, are constrained and enabled by the presence of other factors. Data that combine disparate scales reveal this expansion and contraction of explanatory space.

Consider what makes someone smoke tobacco. We all know that in the United States, fewer people smoke today than fifty years ago, just before a major cultural and regulatory shift began. What the sociologist Jason Boardman and his colleagues have now shown is that whether someone smokes is more heavily influenced by their genes today than it was before smoking was stigmatized. In the 1960s, when every hostess had an ashtray and every stranger had a light, it didn't take much to decide to light up; today, when nicotine prompts dirty looks, it often takes a powerful biological urge that afflicts us unequally. The changed culture makes room for our genes to determine whether we smoke; our genes limit the room for the cultural shift to change what we do. Data only on genes or only on the shift in norms would give us one kind of answer about why people smoke, but both together show us how each constrains the other.

Or consider the rise of antibiotic-resistant staph bacteria. The epidemiologist Diane Lauderdale and her colleagues are analyzing a particular cause of this deadly epidemic in Chicago: prison. Their work triangulates knowledge at the micro scale of how the bacterium passes from one person's skin to another's with knowledge at the macro scale that determines whose skin touches whose: how people move in and out of crowded jails, where they live when they leave, what sports they play. The result is increasingly realistic models of interaction between microbes and humans—not only as individuals or as populations but both at the same time.

This is the future of the population sciences: zooming simultaneously inside individuals, to their microbes or their genes, and outside them to their social norms, their neighborhoods, the laws regulating them. Data that zoom in both directions don't just let us ask new questions—they let us ask a new *type* of question, one that embraces the contingent and contextual nature of human behavior. The social sciences vacillate between broad generalities (which usually turn out to be less general than they appear) and particularistic studies of specific settings. Only data that link the levels of human experience let us fill in the gloriously contingent middle. In the human sciences, the scope conditions are the story.

When you're making a map, you don't just want to know what goes inside the borders; you want to know where the borders are. Explanations should map the space of possibilities, and data that span the levels of human variation let us explore the borders.

CHALLENGING THE VALUE OF A UNIVERSITY EDUCATION

STEVE FULLER

Philosopher; Auguste Comte Chair in Social Epistemology, University of Warwick, U.K.; co-author (with Veronika Lipinska), *The Proactionary Imperative: A Foundation for Transhumanism*

Just in time for the start of the 2015–16 academic year, the U.K. branch of one of the world's leading accounting firms, Ernst & Young, announced that it would no longer require a university degree as a condition of employment. Instead it would administer its own tests to prospective junior employees. In the future, this event will be seen as the tipping point toward the end of the university as an all-purpose credentials mill that feeds the "knowledge-based" economy.

University heads have long complained that economists demean their institution when they reduce its value to a labor-market signal: A good degree = a good job prospect. Yet it would seem that even the economists have been too generous to universities. To be sure, Silicon Valley and its emulators have long administered their own in-house tests to job candidates, but Ernst & Young gained international headlines for being a large mainstream elite employer that has felt compelled to turn to such an approach.

When one considers the massive public and, increasingly, private resources dedicated to funding universities, and the fact that both teaching and research at advanced levels can be and have been done more efficiently outside of universities, the social function of universities can no longer be taken for granted.

As the Ernst & Young story suggests, a prime suspect is the examination system, which has always sat uneasily between the teaching and research functions of the university. At best, exams capture a student's ability to provide a snapshot of a field in motion. But photography is a medium better suited for the dead or the immortal than for ongoing inquiry, where a premium is placed on the prospect that many of our future beliefs will be substantially different from our present ones. A recurring theme in the life stories of great innovators of the modern period, starting with Einstein, is the failure of the exam system to bring out their true capacities. It's not that the thinking of these innovators wasn't transformed by their academic experience; rather, it's that academia lacked an adequate means of registering that transformation.

One charitable but no less plausible diagnosis of many of the errors routinely picked up by examiners is that they result from students' having suspended conventional assumptions in the field in which they are being examined. Yet these assumptions may be challenged if not overturned in the not-too-distant future. Thus, what strikes the examiner as corner-cutting sloppiness may capture an intuition that amounts to a better grasp of the truth of some matter.

But what sort of examination system would vindicate this charitable reading of error and thereby aid in spotting the next generation of innovators? It is not obvious that an in-house exam administered by, say, Ernst & Young will be any less of an epistemic snapshot than an academic exam, if it simply tests for the ability to solve normal puzzles in normal ways. The in-house exam will just be more content-relevant to the employer.

An alternative would be to make all university examinations tests in counterfactual reasoning. In effect, students would be provided access to the field's current state of knowledge—the sort of thing they would normally regurgitate as exam answers—and then be asked to respond to scenarios in which the assumptions behind the answers are suspended in various ways. Thus, students would be tested at once for their sense of how the current state of knowledge hangs together and their ability to reassemble that knowledge strategically under a state of induced uncertainty.

When the great Prussian philosopher-administrator Wilhelm von Humboldt made the "unity of teaching and research" the hallmark of the modern university 200 years ago, his aim was to propel Germany onto the world stage at a time when it was playing catch-up with the political and economic innovations coming from France and Britain. In the process, he transformed the academic into a heroic figure who led by example. "Humboldtian" academics were people whose classroom performance inspired a questing spirit in students as they tried to bring together the often inchoate elements of their field into a coherent whole that pointed the way forward. The ultimate validity of any such synthesis mattered less than the turn of mind the performance represented—one that remained "never at rest," to recall the title of the standard biography of Isaac Newton.

The move by Ernst & Young to administer its own purpose-built examinations is an attempt to produce a more targeted and less expensive version of what it, and much of society, thinks is the source of value in a university education. Universities will fail if they try to compete on those terms. However, they may survive if they learn how to exam in the spirit of Humboldt.

THE HERMENEUTIC HYPERCYCLE

MAXIMILIAN SCHICH

Associate professor for art and technology, University of Texas at Dallas

The most exciting news in our scientific quest to understand the nature of culture is not a single result but a fundamental change in the metabolism of research: With increasing availability of cultural data, increasingly robust quantification nurtures further qualitative insight; taken together, the results inspire new conceptual and mathematical models, which in turn put into question and allow for faster modification of existing data models. Closing the loop, better models lead to more efficient collection of even more cultural data. In short, the hermeneutic circle is replaced by a hermeneutic hypercycle. Driven by the quantification of nonintuitive dynamics, cultural science is accelerated in an autocatalytic manner.

The original "hermeneutic circle" characterizes the iterative research process of the humanist to understand a text or an artwork. The circle of hermeneutic interpretation arises as understanding specific observations presupposes an understanding of the underlying worldview, while understanding the worldview presupposes an understanding of specific observations. As such, the hermeneutic circle is a philosophical concept that functions as a core principle of individual research in the arts and humanities. Friedrich Ast explained it implicitly in 1808; Heidegger and Gadamer further clarified it in the mid-20th century.

Unfortunately the advent of large database projects in the

arts and humanities has almost disconnected the hermeneutic circle in practice. Over decades, database models to embody the underlying worldview were mostly established using formal logic and *a-priori* expert intuition. Database curators were subsequently used to collect vast numbers of specific observations, enabling further traditional research while failing to feed back systematic updates into the underlying database models.

As a consequence, "conceptual reference models" are frozen, sometimes as ISO standards, and out of sync with the nonintuitive complex patterns that would emerge from large numbers of specific observations by quantitative measurement. A systematic data science of art and culture is now closing the loop using quantification, computation, and visualization in addition.

The "hermeneutic hypercycle" is a term that returned no result in search engines before this contribution went online. A product of horizontal meme-transfer, it combines the hermeneutic circle with the concept of the catalytic hypercycle, as introduced by Eigen and Schuster. Like the carbon cycle that keeps our Sun shining and the citric-acid cycle that generates energy in our cells, the hermeneutic circle in data-driven cultural analysis can be understood as a cycle of reactions, here to nurture our understanding of art and culture.

The cycle of reactions is a catalytic hypercycle, as data collection, quantification, interpretation, and data modeling all feed back to catalyze themselves. Their cyclical connection provides a mutual corrective of bias (avoiding an error catastrophe) and leads to a vigorous growth of the field (as we learn what to learn next). In simple words, data collection leads to more data collection, quantification leads to more quantification, interpretation leads to more interpretation, and modeling leads to more modeling. Altogether, data collection nurtures quantification and interpretation, which in turn nurtures modeling, which again nurtures data collection, etc.

It is fascinating to observe the resulting vigorous growth of cultural research. While the naming game of competing terms such as digital humanities, culture analytics, culturomics, or cultural data science is still going on, it becomes ever clearer that we are on our way to a sort of systems biology of cultural interaction, cultural pathways, and cultural dynamics, broadly defined. The resulting "systematic science of the nature of culture" is exciting news, as most issues, from religious fundamentalism to climate change, require cultural solutions and "Nature cannot be fooled."

RETHINKING AUTHORITY WITH THE BLOCKCHAIN CRYPTO ENLIGHTENMENT

MELANIE SWAN

Philosopher, economic theorist, New School for Social Research

If a central problem in the contemporary world could be defined, it might be called adapting ourselves to algorithmic reality. The world is marked by an increasing presence of technology, and the key question is whether we will have an empowering or enslaved relation with it. To have an enabling relation, we may need to mature and grow in new ways. The fear is that just as human-based institutions can oppress, so too might technology-orchestrated realities, and in fact this case might be worse. Blockchain technology is the newest and most emphatic example of algorithmic reality, news that makes us consider our relation with technology more seriously.

Blockchain technology (the secure distributed ledger software that underlies cryptocurrencies like Bitcoin) connotes Internet II (the transfer of value) as a clear successor position to Internet I (the transfer of information). This means that all human interaction regarding the transfer of value—including money, property, assets, obligations, and contracts—could be instantiated in blockchains for quicker, easier, less costly, less risky, and more auditable execution. Blockchains could be a tracking register and inventory of all the world's cash and assets. Orchestrating and moving assets with blockchains concerns both immediate and future transfer, whereby entire classes of industries, like the mortgage industry, might be outsourced to blockchain-based smart contracts, in an even more profound move to the automation economy. Smart contracts are radical as an implementation of self-operating artificial intelligence and also through their parameter of blocktime—rendering time, too, assignable rather than fixed.

Blockchains thus qualify as the most important kind of news, the news of *the new*—something that changes our thinking, causes us to pause, where in a moment we instantly sense that now things might be forever different.

Singularity-class science and technology breakthroughs are news of the new. Advances such as deep learning, self-assembling nanotechnology, 3D-printed synthetic biological life, the genome and the connectome, immersive virtual reality, and self-driving cars might well revolutionize our lives. However impressive, though, these are context-specific technologies. So when a revolution comes along as general and pervasive as to possibly reorchestrate all of the patterns of life, *that* is noticeable news of the new—and that is blockchains.

In existing as a new general class of thing, blockchains reconfigure the definition of what it is to be new. We see the elegance of Occam's razor in redefining what it is to be new. The medium and the message are simultaneously the message; the message is stronger, because it has reconfigured its own form. The blockchain redefines what it is to be new because the medium and message create an entirely new reality and possibility space for how to do things. Where previously there were only hierarchical models for organizing large-scale human activity, blockchains open up the possibility of decentralized models and hybrid models—and not just improved methods but an expansive invitation to new projects. The blockchain news of the new is that we are in a whole new possibility space of diversity and decentralization, no longer confined to hierarchy as the sole mode of coordination.

Blockchains as a class of the new startle us with the discomfort and fear of the unknown, because it's not clear whether blockchain technology is good or bad news. Even more fearful is the implication that it is we who will determine whether the news is good or bad. The underlying technology is, as any other technology, itself neutral—or at least pliable in dual use for both helpful and limiting purposes. Blockchains invite the possibility of action and, more deeply, the responsibility of action. We are developing the sensibility of the algorithmically-aware cryptocitizen, where the expansion of the apparatus of our thought and maturity is proportional to the force of the new experience to be mastered.

Cryptocitizen sensibility and the blockchain news of the new invite the possibility of our greater exercise of freedom and autonomy and rethinking authority. In a Crypto Copernican turn, we can shift the assumed locus of authority from outside ourselves, in external parties, to residing within ourselves. This is the Enlightenment that Kant was after—an advance not just in knowledge but also in authority-taking. A Cambrian explosion of experimentation in new models of economics and governance could ensue. Blockchains, being simultaneously global and local, could coordinate the effort, equitably joining the diversity of every permutation of micropolis with the cohesion of the macropolis, in connected human activity on digital smart-networks.

When we have an encounter with *the new*, that is indeed news. This is the news we crave and seek above all else, an encounter with the sublime that reconfigures our reality. The galvanizing essence of the blockchain news of the new is the possibility of our further expansion into our own potentiality. The Crypto Enlightenment is the kind of authority-taking change we can make within ourselves as a maturation toward the realization of an empowering relationship with technology.

ENVOI: WE MAY ALL DIE HORRIBLY

ROBERT SAPOLSKY

Professor of biology, neurology, and neurosurgery, Stanford
University; author, *Monkeyluv*

Well, the obvious place to start answering the latest *Edge* Question was that little paper about scientists using CRISPR technology to show that *Homo naledi* buried their dead next to coursing rivers on Mars. Despite that slam-dunk of a choice, my vote for the most interesting/important piece of recent science news comes in two parts.

Part I: Back to December 2013, when a one-year-old boy in a Guinean village died an agonizing death, and as a result many of us learned the names of some West African countries for the first time. By now, everyone is familiar with the broad features of the West African Ebola-virus epidemic. The disease had been flaring up intermittently in Central Africa and then quickly burning out. It is devastatingly lethal, rapidly killing most people infected. The virus requires contact with bodily fluid for transmission, and it has evolved to facilitate that, as sufferers die in an explosion of bodily fluids—projectile vomiting, continuous diarrhea, and, with some viral strains, external hemorrhaging. If Joseph Conrad had known about Ebola, he would have written it into *Heart of Darkness*.

And then the virus made the nightmare jump from the rain forest to high-density urban settings, bursting out almost simultaneously in the cities and towns of Sierra Leone, Liberia, and Guinea. This was inevitable, given modern mobility.

The countries were utterly unprepared—poor, with large shantytown populations, still recovering from years of war, and with minimal medical infrastructure (Liberia had some fifty doctors at the time). Nearly 30,000 cases, over 11,000 deaths—numbers that are probably underestimates. An even larger number of people sickened and died in the secondary death toll, because hospitals were overwhelmed with Ebola patients. At the epidemic's peak, there were thousands of new cases each week.

Entire extended families were wiped out, as people cared for relatives, or washed their corpses, as per custom. Villages and towns were emptied, the capitals became ghost towns as governments urged people to stay in their homes, to not touch other people. Any semblance of healthy economic activity disappeared. Hysteria flared in various predictable ways: Some denied the existence of the disease, insisted it was a hoax; quarantine centers were ransacked by crowds intent on "liberating" their relatives; burial teams were attacked when they came for bodies. Virus-free survivors were shunned. And, naturally, suspicion and fear of the disease prompted various shoot-the-messenger scenarios; in a machete-the-messenger variant, eight aid workers were hacked to death in a village in Guinea.

Suspicion and fear also played out in conspiracy theories: A leading Liberian newspaper, a Liberian professor in the U.S., *Pravda*'s English edition, and Louis Farrakhan all declared that Ebola was invented by the West. As a bioweapon that escaped from a lab. As a bioweapon being field-tested on Africans. As a strategy to decrease the African population. Or, in *Pravda*'s capitalist critique, as a bioweapon released to let the West hold the world hostage by charging exorbitant prices for a patented cure—i.e., a military-pharmacology complex. And of course, various West African clergy announced that God had sent Ebola to punish West African countries for the supposed laxity with which they persecuted homosexuals.

Unless you were a conspiracy theorist, the story had no villains (although the World Health Organization got a lot of criticism for not being prepared, as if that were feasible). But there were heroes. Doctors Without Borders achieved cult status, both for its presence and effectiveness but also for its lack of institutional BS. Medical missionaries were extraordinary (something I was initially loath to admit; having worked regularly in Africa for thirty years, I spout a secular leftist's obligatory condemnation of missionaries). And most of all, there were the heroics of the West African healthcare workers—doctors, nurses, ambulance drivers, burial teams—with scant resources to keep them effective, or safe. They accounted for a tenth of the deaths in the epidemic. In August 2014, *Science* published an Ebola paper five of whose West African authors were already dead.

We watched all this from afar, free to turn the page to the next story. And then it came home.

First were the handful of infected expatriate healthcare workers who were brought, using remarkable care and containment methods, back to the West for almost invariably successful treatment. Then came Thomas Eric Duncan, a Liberian visiting relatives in Dallas, whose Ebola was the first case diagnosed in the U.S. He was a good Samaritan, having driven a sick pregnant neighbor to the hospital thinking her illness was pregnancy-related. Thus he had failed to note his exposure on health forms and was allowed to fly; he went to an ER in Dallas with beginning symptoms of the disease, and someone failed to ask him whether he had come from another country, let alone from the

West African hot zone. He was sent home with an antibiotic prescription and returned to the hospital a few days later to die. Soon two of the nurses who had cared for him contracted the disease. The second, aware of the first's diagnosis, nevertheless flew to Akron to buy a wedding dress; the shop went out of business as a result (which did not prevent her, in a move that defines *chutzpah*, from requesting a refund on her ostensibly tainted purchase).

Then came the doctor back from treating Ebola in West Africa who visited a bowling alley in Brooklyn the evening before he developed symptoms. And the returning nurse who either did or didn't represent a health risk but refused to be quarantined, biking around her Maine town for photographers, asymptomatic and virus-free, wearing her bike helmet.

By then, we were all collectively wetting our pants with terror. The question that usually appeared at the end of past articles about obscure Central African hemorrhagic viruses—"Could it happen here?"—had become "It's happening here, isn't it?" And this produced what I consider the most important scientific moment in 2014. It wasn't spurred by Legionnaire's disease, toxic shock syndrome, the anthrax scare, or even AIDS—and certainly not by global warming. The U.S. population finally reached the collective realization that we may all die horribly unless some scientists figure out a way to save us.

Part II: The (beginning of a) resolution: As the most scientifically significant moment of 2015, a twenty-eight-author team published in *Lancet* the results of a Phase II clinical trial of an Ebola vaccine.[*] The carefully designed trial involved nearly 8,000 Guineans and demonstrated the vaccine's 100-percent ef-

---

[*] http://dx.doi.org/10.1016/S0140–6736(15)61117–5

fectiveness at preventing disease occurrence when administered immediately after exposure to someone with Ebola.

Yes, this isn't the end of Ebola disease (and the research started long before the West African epidemic). But it's a rough approximation of how science can indeed act to save us. It would be nice if the general public thought the same.

BOOKS BY JOHN BROCKMAN

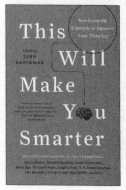

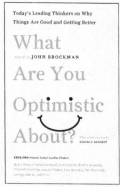